Response Surface Methods - Theory, Applications and Optimization Techniques

Edited by Valter Silva
and João Sousa Cardoso

Published in London, United Kingdom

Response Surface Methods - Theory, Applications and Optimization Techniques
http://dx.doi.org/10.5772/intechopen.1001640
Edited by Valter Silva and João Sousa Cardoso

Contributors
Audberto Reyes-Rosas, Babatunde L. Adeleke, Daniella-Esperanza Pacheco-Catalán, Gabriel O. Adebayo, Hembadoon R. Asoo, Joseph S. Alakali, Julius K. Ikya, Kaizhe (Kaiser) Qiu, Kazeem A. Osuolale, Mehmet Fatih Demiral, Mohammed I. Yusufu, Oyebola Odunayo Olabinjo, Prócoro Gamero-Melo, Sasirot Khamkure, Shaoping Qiu, Sofía-Esperanza Garrido-Hoyos, Victoria Bustos-Terrones

First published in London, United Kingdom, 2024 by IntechOpen
IntechOpen is the global imprint of INTECHOPEN LIMITED, registered in England and Wales, registration number: 11086078, 167-169 Great Portland Street, London, W1W 5PF, United Kingdom

British Library Cataloguing-in-Publication Data
A catalogue record for this book is available from the British Library

Additional hard and PDF copies can be obtained from orders@intechopen.com

Response Surface Methods - Theory, Applications and Optimization Techniques
Edited by Valter Silva and João Sousa Cardoso
p. cm.
Print ISBN 978-0-85466-767-3
Online ISBN 978-0-85466-766-6
eBook (PDF) ISBN 978-0-85466-768-0

Meet the editors

Valter Silva is a senior researcher in environment and energy at the University of Aveiro, Portugal. He obtained his Ph.D. in Chemical and Biological Engineering from the University of Porto in 2009. Since 2012, he has led a research team devoted to the application of experimental and numerical solutions on environmental and energy topics. In the last five years, Dr. Silva has coordinated national and international projects with leading universities across the world, including the Massachusetts Institute of Technology (MIT) and Carnegie Mellon, NY, raising approximately $2.5 million in funds. He has supervised more than thirty students. He has participated in more than twenty-five national and international projects in energy and environment and has authored two books, eleven book chapters, and more than sixty papers in international peer-reviewed journals.

João Cardoso obtained a Ph.D. in Mechanical Engineering from the Instituto Superior Técnico, University of Lisbon, Portugal. He is currently operating as a development and system engineer and project manager in the aerospace sector with a strong foundation in mechanical engineering, industrial processes, and renewable energies. He is responsible for the thermal development of the Comet Interceptor and FORUM missions by the European Space Agency (ESA). Dr. João's journey has taken him through thermal hardware development, project management, process optimization, simulation, and techno-economic and risk analysis. He has around forty influential works to his credit, including books, book chapters, and high-impact scientific papers in international journals.

Contents

Preface

Response Surface Methods – Theory, Applications and Optimization Techniques is a comprehensive collection of chapters delving into the multifaceted domain of Response Surface Methodology (RSM). This book is a result of the collective expertise and contributions of numerous distinguished authors. It is a valuable resource for researchers, engineers, and scientists looking to collect RSM's power, thus enhancing their experimental designs and optimization processes.

RSM is a robust tool for designing experiments, analyzing experimental data, and optimizing processes. Though it was initially developed for optimizing industrial processes, RSM has expanded its action and utility across multiple disciplines such as engineering, chemistry, biology, health care, and environmental sciences. The adaptability of RSM lies in its ability to model and analyze complex interactions within experimental data, providing insightful solutions for process optimization and quality improvement.

The book is divided into two sections, each focusing on different aspects of RSM, from fundamental theories to advanced applications.

Section I introduces RSM with a foundational understanding of the technique, tracing its historical development and establishing key concepts and definitions. Section II discusses optimization techniques and advanced topics associated with RSM.

This volume has been carefully curated to provide broadness and depth to the subject of RSM, reaching a wide audience from beginners to advanced practitioners. The contributing authors' collaborative efforts not only cover the theoretical aspects of RSM but also provide practical insights and applications, making the book a valuable reference for anyone looking to leverage RSM in their work.

The editorial team hopes that *Response Surface Methods – Theory, Applications and Optimization Techniques* will inspire and guide researchers and professionals through their endeavors, helping them to achieve greater efficiency, precision, and innovation in their experimental designs and optimization projects.

Valter Silva
Centre for Environmental and Marine Studies (CESAM),
University of Aveiro,
Aveiro, Portugal

João Sousa Cardoso
Instituto Superior Técnico,
University of Lisbon,
Lisbon, Portugal

Section 1

Introduction to Response Surface Methods and Theoretical Foundations

Chapter 1

Historical Background of RSM

Hembadoon R. Asoo, Joseph S. Alakali, Julius K. Ikya and Mohammed I. Yusufu

Abstract

The historical background of response surface methodology (RSM) traces its evolution from early experimental design principles to its widespread adoption in various industrial applications. This paper examines the development of experimental design techniques, initial approaches to optimization, and the statistical foundation underlying RSM. It explores the pioneering contributions of G. E. P. Box, who played a pivotal role in advancing RSM. The evolution of RSM terminology and its integration with computer technology are discussed, along with challenges and criticisms encountered over time. The cross-disciplinary adoption of RSM is highlighted, emphasizing its relevance across diverse fields. Modern developments and innovations in RSM are examined, including advancements in modeling techniques and optimization algorithms. The limitations of RSM, such as assumptions of polynomial models and sensitivity to initial experimental design, are acknowledged, with strategies proposed for overcoming these challenges. Overall, this abstract provides a comprehensive overview of the historical trajectory, industrial significance, and contemporary advancements of RSM, offering insights into its application and potential for future research.

Keywords: response surface methodology, experimental design, optimization, G. E. P. Box, computer technology, cross-disciplinary adoption

1. Introduction

Response surface methodology (RSM) is a mid-twentieth-century statistical tool for optimizing processes and understanding complex relationships between variables. It was pioneered by statisticians George E. P. Box and K. B. Wilson and has gained prominence in diverse fields. Its roots trace back to the design of experiments and statistical modeling, with notable contributions in engineering, chemistry, and manufacturing. RSM has evolved over time into a powerful technique for efficiently exploring and improving processes by systematically adjusting variables. It has become integral in experimental design, offering a structured approach to achieve optimal outcomes and enhance productivity in different fields.

1.1 Development of experimental design

The evolution of experimental design principles leading to the formulation of response surface methodology (RSM) is a rich history that spans statistical innovations and practical applications. The journey begins with the foundational work of Sir Ronald A. Fisher in the early twentieth century.

Fisher's groundbreaking contributions introduced the concept of randomized experiments and factorial designs, emphasizing the importance of randomization to reduce bias and the simultaneous study of multiple factors. His work laid the groundwork for modern experimental design, allowing researchers to explore efficiently the effects of various factors on a response variable [1].

The subsequent development of analysis of variance (ANOVA) by Fisher in the 1920s became a pivotal statistical tool in understanding the sources of variability in experimental data. ANOVA allowed researchers to systematically partition total variation into different components attributable to different factors, providing a rigorous approach to statistical analysis [1].

In the mid-twentieth century, George E. P. Box and others extended Fisher's work by introducing central composite designs (CCD). CCD combined the efficiency of factorial designs with the exploratory power of response surface designs. This marked a significant step toward optimizing processes, allowing researchers to efficiently explore the experimental space and identify optimal conditions [2].

Box and Wilson's contributions paved the way for the development of Box-Behnken Designs in the 1960s. These designs, named after George E. P. Box and Donald Behnken, offered a more resource-efficient approach. Box-Behnken Designs required fewer experimental runs while still providing accurate estimates of response surfaces, making them particularly valuable in practical applications with limited resources [3].

The formalization of response surface methodology (RSM) in the late twentieth century represented a culmination of these advancements. RSM integrates statistical and mathematical techniques to model and optimize complex processes with multiple variables. It employs mathematical models to represent the relationships between input variables and the response of interest, allowing for the identification of optimal conditions [4].

In conclusion, the evolution of experimental design principles leading to RSM reflects a continuous refinement aimed at maximizing the efficiency of experimentation and optimization processes. From Fisher's foundational work on randomized experiments to the integration of statistical models in RSM, each step has contributed to a more systematic and sophisticated approach to experimental design in scientific research and industrial applications.

1.2 Early approaches to optimization

Early approaches to optimization laid the foundation for the development of response surface methodology (RSM) as a powerful tool in the field. These methods, rooted in mathematical and statistical principles, paved the way for more sophisticated optimization techniques. Some key early approaches that influenced the evolution of optimization methodologies are presented below.

2. Classical optimization techniques

Calculus-based methods: The roots of optimization can be traced back to calculus-based techniques, such as the method of steepest descent and Newton's method. These methods focus on finding extrema by analyzing the slope or curvature of a function.

Lagrange multipliers: Introduced by Joseph-Louis Lagrange in the 18th century, Lagrange multipliers extended optimization to constrained problems, allowing researchers to optimize subject to certain conditions.

3. Experimental design

Factorial designs: Early experimental design techniques, particularly factorial designs, played a crucial role. Developed by Sir Ronald A. Fisher in the 1920s, factorial designs enabled researchers to study the effects of multiple factors simultaneously, providing a structured approach to experimentation.

Fractional Factorial Designs: Building upon factorial designs, fractional factorial designs were developed to efficiently explore factor interactions with fewer experimental runs. This approach proved valuable for optimization in situations where resources were limited.

4. Response surface methodology (RSM)

Introduction of RSM: The formalization of response surface methodology is often credited to George E. P. Box and K. B. Wilson [2] in the mid-20th century. RSM emphasizes modeling and analyzing the relationship between input variables and the output response of a system.

Central Composite Designs (CCD): Within RSM, CCDs, introduced to efficiently explore the response surface with a combination of factorial and axial points enhanced the precision of estimating key parameters, making optimization more effective.

5. Evolutionary algorithms

Genetic algorithms: Inspired by the process of natural selection, genetic algorithms emerged in the 1960s and 1970s. Developed by John Holland [5], these algorithms involve the evolution of a population of potential solutions through crossover, mutation, and selection. Genetic algorithms demonstrated effectiveness in solving complex optimization problems.

6. Stochastic optimization

Simulated annealing: In the 1980s, simulated annealing, inspired by the annealing process in metallurgy, was introduced as a stochastic optimization technique that explores the solution space by allowing occasional uphill movements, thus enabling escape from local optima.

Early optimization techniques, rooted in calculus, experimental design, and statistical methodologies, laid the groundwork for the development of response surface methodology. RSM, with its emphasis on modeling and analyzing complex systems, became a powerful optimization tool. The evolution from classical methods to advanced techniques like genetic algorithms and simulated annealing reflects the continuous refinement of optimization methodologies over the years.

6.1 Statistical foundations

Response surface methodology (RSM) is deeply rooted in statistical principles, drawing on key concepts and methods that have significantly influenced its creation. The development of RSM was driven by the need to optimize complex processes efficiently. Below are statistical foundations that played a pivotal role in shaping the methodology.

6.1.1 Design of experiments (DOE)

Factorial Designs: The foundation of RSM lies in factorial designs, a concept introduced by Sir Ronald A. Fisher in the 1920s. Factorial designs allow researchers to study the effects of multiple factors simultaneously. This principle is fundamental to RSM, which often employs factorial designs to efficiently explore the experimental space.

Fractional Factorial Designs: Extending factorial designs, fractional factorial designs were crucial in the development of RSM. They enable the exploration of factor interactions with fewer experimental runs, making it feasible to study a large number of factors and their combinations.

6.1.2 Regression analysis

Multiple regression: RSM heavily relies on multiple regression analysis to model the relationship between input variables and the response of a system. Multiple regression, a statistical technique dating back to the early 20th century, allows for the identification of significant factors and their interactions.

6.1.3 Central composite designs (CCD)

Optimal experimental design: The concept of optimal experimental design, particularly central composite designs, played a pivotal role in refining RSM. Box and Wilson [1], in their influential work, emphasized the importance of optimizing the locations of experimental points. CCDs efficiently combine factorial and axial points to provide precise estimates of key parameters.

6.1.4 Statistical hypothesis testing

Analysis of variance (ANOVA): The application of analysis of variance, a statistical method developed by Fisher [6], is integral to RSM. ANOVA enables the decomposition of total variance into components attributable to different sources, aiding in the assessment of the significance of factors and interactions.

6.1.5 Statistical modeling

Response surface modeling: The central idea behind RSM is the construction of response surface models. These models, built through statistical techniques, capture the relationship between input variables and the response. They provide a comprehensive understanding of the system's behavior and guide the optimization process.

Optimization Criteria: Statistical criteria for optimization, such as the desirability function, are employed within the RSM framework. These criteria help researchers define and achieve the optimal conditions for a process by considering multiple response variables simultaneously.

6.1.6 Robust parameter design

Taguchi Methods: While not directly forming the basis of RSM, the Taguchi methods, developed by Taguchi [7], contributed to the robust parameter design aspect. The methods focus on minimizing variability and optimizing performance in the presence of noise factors, aligning with the goals of RSM in practical applications.

The statistical foundations of RSM encompass a rich array of methodologies, including factorial and fractional factorial designs, regression analysis, optimal experimental design, hypothesis testing, response surface modeling, and robust parameter design. The integration of these statistical principles has elevated RSM to a powerful and versatile tool for optimization in various fields.

7. Industrial applications of RSM

The historical context of RSM is deeply intertwined with the need for efficient optimization of processes to enhance product quality and operational efficiency.

7.1 Early applications in industry

The initial application of RSM in industrial settings traces back to the pioneering work of George E. P. Box and K. B. Wilson [2]. In the early 1950s, these researchers introduced the methodology as a systematic approach to optimizing complex processes and improving product quality.

7.2 Chemical and manufacturing industries

Chemical engineering: One of the earliest industrial applications of RSM was in chemical engineering. Researchers and engineers in this field embraced RSM to optimize reaction conditions, maximize yield, and minimize variability in chemical processes.

Manufacturing: RSM found applications in optimizing production processes, reducing defects, and improving the overall quality of manufactured goods in the manufacturing field. The methodology's ability to navigate complex parameter spaces and identify optimal conditions became instrumental in enhancing efficiency.

7.3 Food and pharmaceutical industries

Food Processing: In the food industry, factors such as temperature, pressure, and ingredient proportions significantly affect product quality, and RSM plays a key role [8]. Researchers utilized RSM to optimize these factors, leading to improved taste, texture, and shelf life of food products.

Pharmaceuticals: In the pharmaceutical industry, RSM found applications in optimizing drug formulations and manufacturing processes. By systematically exploring the effects of various factors, researchers could enhance drug quality, improve yields, and reduce costs.

7.4 Engineering and technology

Process Engineering: RSM was widely adopted in process engineering to optimize parameters in the design and operation of various systems. From optimizing the efficiency of energy production to improving the performance of mechanical systems, RSM proved invaluable [9].

Product Design: In engineering design, RSM contributed to optimizing product performance and reliability. It became a valuable tool in the design of experiments to understand how different factors influence the final product and how to achieve desired outcomes efficiently.

7.5 Quality improvement initiatives

Six Sigma: RSM became an integral part of Six Sigma methodologies, which focus on minimizing variability and defects in processes. By employing RSM, Six Sigma practitioners could identify optimal process conditions that led to a reduction in defects and an improvement in overall quality.

Lean Manufacturing: RSM complemented lean manufacturing principles by providing a structured approach to process optimization. It allowed organizations to identify critical process parameters and optimize them to achieve lean objectives, such as reducing waste and improving efficiency.

7.6 Continued impact and evolution

Over the years, RSM has evolved with advancements in statistical techniques and computational tools. Its impact on industrial settings continues to grow as industries increasingly recognize the value of systematic optimization in achieving operational excellence.

Today, RSM is applied not only in traditional manufacturing but also in emerging fields such as biotechnology, nanotechnology, and information technology, showcasing its adaptability and enduring relevance.

The historical context of RSM's application in industrial settings reflects a journey of continuous improvement and innovation. From its early roots in chemical engineering to its widespread adoption across diverse industries, RSM has had a profound impact on optimizing processes and enhancing product quality. As industries continue to embrace data-driven decision-making, the legacy of RSM as a valuable optimization tool persists.

8. Contributions of George E. P. Box to response surface methodology

George E. P. Box, a prominent statistician, made significant and lasting contributions to the development and popularization of response surface methodology (RSM). His work, both theoretical and applied, has had a profound impact on the field of experimental design and optimization. Some of the key contributions to RSM are highlighted below.

8.1 Introduction and formalization of RSM

In collaboration with Wilson, Box introduced RSM in the early 1950s as a systematic and efficient approach to optimizing complex processes. Their seminal paper, "On the Experimental Attainment of Optimum Conditions" [2], laid the foundation for RSM, emphasizing the importance of statistical experimental design in the pursuit of optimal conditions.

8.2 Central composite designs (CCD)

Box [10] played a pivotal role in the development of central composite designs (CCD), a key aspect of RSM. CCDs are a type of experimental design that combines factorial points with axial points to explore the experimental space efficiently. The optimal arrangement of experimental points, as advocated by Box, enhances the precision of estimating key parameters in the response surface model.

8.3 Emphasis on optimal experimental design

Box, a strong advocate for optimal experimental design, stressed the importance of carefully selecting experimental conditions to yield precise and reliable results. His work highlighted the significance of efficiently using experimental resources by strategically placing experimental points in the design space.

8.4 Robust parameter design

Box's contributions extended to the concept of robust parameter design, emphasizing the need to make processes less sensitive to uncontrollable factors. In collaboration with Dr. Norman R. Draper [11], he introduced the concept of robustness in parameter design, providing a statistical framework to achieve products and processes that are less sensitive to variations.

8.5 Extension of RSM beyond factorial designs

Box recognized the limitations of traditional factorial designs and extended the application of RSM to non-factorial designs. This broader perspective allowed the application of RSM to a wide range of experimental scenarios, enabling researchers to study and optimize complex systems with multiple variables [12].

8.6 Development of response surface models

Box contributed significantly to the development of response surface models, which are mathematical representations of the relationship between input variables and the response of a system. His work emphasized the importance of fitting accurate models to experimental data to gain insights into the underlying processes.

8.7 Practical applications and collaboration

Box's contributions to RSM were not confined to theoretical advancements. He actively collaborated with researchers from various fields, applying RSM to practical problems in industries such as chemistry, engineering, and manufacturing. This practical application of RSM helped establish its credibility and usefulness in real-world scenarios.

8.8 Legacy and recognition

Box's contributions to RSM have left an enduring legacy. His influential book, "Statistics for Experimenters" [12], co-authored with Norman R. Draper, remains a foundational resource in the field. Box's work has been widely recognized, and he received numerous awards and honors, including the Shewhart Medal and the Deming Medal [13].

George E. P. Box played a pioneering role in the development and popularization of response surface methodology. His contributions have shaped the way researchers approach experimentation and optimization. His work has had a lasting impact on both the theoretical foundations and practical applications of RSM, solidifying his legacy as a key figure in the history of statistical experimental design.

9. Evolution of RSM terminology

The evolution of response surface methodology (RSM) terminology has undergone a significant journey, reflecting both the theoretical advancements and practical applications of the methodology. From its early conceptualization to its widespread adoption in various fields, the standardization of RSM terminology has played a crucial role in facilitating communication and understanding among researchers.

9.1 Early terminology and conceptualization*

The early development of RSM terminology can be traced back to the seminal work of George E. P. Box and K. B. Wilson in the 1950s. In their foundational paper, "On the Experimental Attainment of Optimum Conditions" [2], terminologies such as "response surface" and "optimum conditions" were introduced, laying the groundwork for RSM.

The terminology in these early stages primarily revolved around experimental design principles and the fitting of response surface models to data. Concepts such as "factorial designs," "central composite designs," and "regression modeling" formed the basis of RSM terminology [2].

9.2 Standardization and formalization

As RSM gained popularity and acceptance in various fields, efforts were made to standardize and formalize its terminology. Textbooks and reference materials began to emerge, providing a common language for researchers and practitioners.

The influential book *Response Surface Methodology* by Raymond H. Myers, Douglas C. Montgomery, and Christine M. Anderson-Cook [9] played a significant role in standardizing RSM terminology. Terms such as "design matrix," "response variable," "regression coefficients," and "optimization criteria" became widely accepted within the RSM community.

9.3 Extension and refinement

With advancements in statistical theory and computational tools, the terminology of RSM continued to evolve to encompass a broader range of concepts and methodologies. Terms related to robust parameter design, sensitivity analysis, and model validation were introduced to address emerging challenges and opportunities.

The incorporation of concepts from related disciplines, such as machine learning and optimization, also influenced the evolution of RSM terminology. Terms like "meta-modeling," "surrogate models," and "global optimization" became increasingly relevant as RSM expanded its scope beyond traditional experimental design.

9.4 Integration with quality improvement initiatives

The integration of RSM with quality improvement initiatives such as Six Sigma and Lean Manufacturing led to the adoption of additional terminology. Terms like "defect reduction," "process capability," and "continuous improvement" became part of the RSM lexicon, reflecting its application in enhancing product quality and operational efficiency [8].

9.5 Contemporary trends and future directions

In recent years, contemporary trends such as big data analytics, artificial intelligence, and Bayesian methods have influenced the evolution of RSM terminology. Terms like "data-driven modeling," "machine learning algorithms," and "Bayesian optimization" are increasingly being integrated into the RSM framework.

The evolution of RSM terminology is ongoing and reflects the dynamic nature of the field and its adaptation to new challenges and opportunities in research and practice. As RSM continues to evolve, the standardization and refinement of its terminology will remain essential for effective communication and collaboration among researchers and practitioners.

Response surface methodology terminology has been characterized by a progression from early conceptualization to standardization, formalization, and integration with related disciplines and quality improvement initiatives. The standardization of RSM terminology has played a crucial role in facilitating communication and understanding and enables researchers and practitioners to effectively apply RSM principles to solve complex problems in various fields.

10. Integration of computer technology in RSM

The integration of computer technology into response surface methodology (RSM) has revolutionized the way researchers design experiments, analyze data, and optimize processes. This historical integration has significantly enhanced the efficiency, accuracy, and scope of RSM applications. Some of the key advancements that facilitated the integration of computer technology into RSM over time are discussed below.

10.1 Computational simulation and modeling

Early integration: In the initial stages, the integration of computers into RSM enabled researchers to perform computational simulations and develop mathematical models to describe complex systems. This facilitated the exploration of response surfaces and the identification of optimal process conditions [9].

Advanced modeling techniques: With the advancement of computational capabilities, more sophisticated modeling techniques, such as neural networks, support vector machines, and ensemble methods, have been integrated into RSM. These techniques allow for the development of highly accurate and flexible response surface models, particularly in situations with nonlinear and high-dimensional data [14].

10.2 Design and analysis of experiments

Design optimization algorithms: Computer technology has enabled the development of efficient algorithms for optimizing experimental designs in RSM. These algorithms can handle complex design spaces, incorporate constraints, and generate optimal designs tailored to specific objectives. Examples include genetic algorithms, simulated annealing, and response surface methods for design optimization [15].

Statistical software packages: The availability of user-friendly statistical software packages, such as JMP, Minitab, and R, has facilitated the design and analysis of RSM experiments. These packages offer a wide range of tools for experimental design, regression modeling, model diagnostics, and optimization, making RSM more accessible to researchers and practitioners [9].

10.3 Visualization and interpretation of results

Graphical visualization tools: Computer technology has enabled the development of graphical visualization tools for exploring response surfaces and analyzing experimental results. Visualization techniques such as contour plots, 3D surface plots, and sensitivity analyses help researchers gain insights into the relationships between input variables and responses, facilitating interpretation and decision-making [8].

Interactive dashboards: Recent advancements in web-based technologies have led to the development of interactive dashboards for RSM analysis. These dashboards allow users to dynamically explore response surfaces, adjust input variables, and visualize the impact on responses in real time, enhancing the interactive and exploratory nature of RSM analysis [16].

10.4 High-performance computing

Parallel computing: The integration of high-performance computing (HPC) technologies has accelerated the computational-intensive tasks involved in RSM, such as model fitting, parameter estimation, and optimization. Parallel computing platforms, including clusters, grids, and cloud computing, enable researchers to leverage massive computational resources for large-scale RSM studies [17].

The integration of computer technology into response surface methodology has transformed the way researchers approach experimental design, data analysis, and process optimization. Key advancements such as computational simulation and modeling, design optimization algorithms, statistical software packages, visualization tools, and high-performance computing have collectively enhanced the efficiency, accuracy, and scope of RSM applications, making it a valuable tool for solving complex problems in various fields.

11. Challenges and criticisms over time

Over time, response surface methodology (RSM) has faced several challenges and criticisms, prompting researchers to adapt and improve the methodology to address these concerns. Following are some of the historical challenges and criticisms faced by RSM and the methodology's evolution in response.

11.1 Limited applicability to nonlinear systems

Challenge: One early criticism of RSM was its limited applicability to nonlinear systems, as traditional RSM techniques were primarily designed for linear models.

Nonlinear relationships between input variables and responses posed challenges in accurately modeling and optimizing complex systems [4].

Response: To address this challenge, researchers developed advanced modeling techniques, such as neural networks, support vector machines, and response surface methods for non-linear models. These techniques expanded the scope of RSM to accommodate nonlinear relationships, enabling more accurate modeling and optimization of complex systems [14].

11.2 Sensitivity to experimental error and noise

Challenge: RSM is sensitive to experimental error and noise, which can affect the accuracy and reliability of response surface models. Variability in experimental conditions, measurement errors, and uncontrolled factors can lead to biased estimates and poor model predictions [8].

Response: Researchers developed robust parameter design techniques within the framework of RSM to address the sensitivity to experimental error and noise. Robust parameter design aims to make processes less sensitive to variations by identifying optimal operating conditions that minimize the impact of uncontrollable factors [11].

11.3 Inadequate experimental design

Challenge: Inadequate experimental design can undermine the effectiveness of RSM by failing to capture important factors and interactions. Poorly designed experiments may result in inefficient use of resources, inaccurate parameter estimates, and unreliable model predictions [4].

Response: Improvement efforts were made for experimental design practices within RSM through the development of optimal design algorithms and guidelines. These advancements helped researchers generate efficient and informative experimental designs tailored to specific objectives, thereby enhancing the reliability and robustness of RSM analyses [15].

11.4 Interpretability and transparency of models

Challenge: Complex response models generated by RSM techniques may lack interpretability and transparency, making it difficult for researchers and practitioners to understand the underlying relationships between input variables and responses. This can hinder decision-making and limit the practical utility of RSM [16].

Response: Researchers have focused on improving the interpretability and transparency of response surface models by incorporating model diagnostics, sensitivity analyses, and visualization techniques. These approaches help identify influential factors, assess model adequacy, and communicate results effectively, enhancing the practical utility of RSM in various applications [9].

Response Surface Methodology has faced challenges and criticisms over time. However, the methodology has evolved in response to feedback, with advancements in modeling techniques, robust parameter design, experimental design optimization, and model interpretability. These improvements have enhanced the effectiveness, accuracy, and practical utility of RSM in addressing complex problems in various fields.

12. Cross-disciplinary adoption of RSM

Response surface methodology (RSM) has expanded beyond its initial domains and has become a valuable tool in a wide range of scientific and engineering disciplines. Its versatility, robustness, and effectiveness in modeling complex systems have led to its adoption across various fields.

12.1 Chemistry and chemical engineering

RSM has found extensive applications in chemistry and chemical engineering for optimizing reaction conditions, improving yields, and enhancing product quality. Researchers use RSM to study the effects of multiple factors on chemical processes, such as temperature, pressure, and catalyst concentration, and optimize process parameters accordingly [2]. The methodology has been instrumental in developing efficient and sustainable chemical processes, reducing waste, and improving the efficiency of resource utilization. Its applications range from pharmaceutical synthesis to food processing and environmental remediation [14].

12.2 Agriculture and agroecology

In agriculture and agroecology, RSM is employed to optimize agricultural practices, enhance crop yields, and improve soil and water management. Researchers use RSM to study the interactions between agronomic factors, environmental conditions, and crop performance, leading to more efficient and sustainable agricultural systems [9]. The application of RSM in precision agriculture helps farmers optimize inputs such as fertilizers, irrigation, and crop protection products, leading to increased productivity and reduced environmental impact. Additionally, the methodology contributes to the development of innovative farming techniques, such as controlled-environment agriculture and hydroponics [8].

12.3 Biotechnology and pharmaceutical sciences

RSM plays a crucial role in biotechnology and pharmaceutical sciences for optimizing bioprocesses, developing drug formulations, and improving the production of biopharmaceuticals. Researchers use the methodology to optimize culture conditions, fermentation parameters, and downstream processing steps, leading to higher yields and improved product quality [9]. It has facilitated the development of new drug delivery systems, dosage forms, and therapeutic products by systematically exploring the effects of formulation variables on drug release kinetics, stability, and bioavailability. The methodology also supports quality-by-design (QbD) approaches in pharmaceutical manufacturing, ensuring product quality and regulatory compliance [14].

12.4 Environmental engineering and sustainability

RSM contributes to environmental engineering and sustainability efforts by optimizing pollution control technologies, waste management strategies, and renewable energy systems. Researchers use RSM to design and optimize processes for wastewater treatment, air pollution control, and solid waste management, aiming to minimize environmental impact and resource consumption [2]. It supports the

DOI: http://dx.doi.org/10.5772/intechopen.1005302

development of sustainable energy systems, such as biofuels, solar cells, and wind turbines, by optimizing materials, process parameters, and system configurations. The methodology also aids in life cycle assessment (LCA) studies, helping assess the environmental impacts of products and processes and to identify opportunities for improvement [8].

Response surface methodology (RSM) has transcended disciplinary boundaries and become a valuable tool in various scientific and engineering disciplines. By providing a systematic framework for experimentation, modeling, and optimization, it contributes to innovation, efficiency, and sustainability across diverse fields, addressing complex challenges and advancing knowledge and technology.

13. Modern development and innovations

In recent years, response surface methodology (RSM) has undergone significant development and innovation, leveraging advancements in computational techniques, data analytics, and interdisciplinary collaboration to address emerging challenges and opportunities. Some of the key recent advancements in RSM include:

13.1 Integration with machine learning and artificial intelligence

Recent developments have seen the integration of RSM with machine learning and artificial intelligence (AI) techniques, enabling the construction of more accurate and robust response surface models. Techniques such as neural networks, support vector machines, and deep learning algorithms complement traditional RSM approaches, particularly in handling nonlinear and high-dimensional data [18].

13.2 Bayesian optimization and sequential design strategies

Bayesian optimization methods and sequential design strategies have emerged as powerful tools for optimizing experimental designs and response surface models. These approaches allow for adaptive experimentation, wherein the design is updated iteratively based on the information gathered from previous experiments, leading to faster convergence to optimal solutions [19].

13.3 High-dimensional optimization and big data analytics

With the proliferation of high-dimensional data in various fields, RSM has evolved to handle complex optimization problems involving a large number of input variables. Advanced algorithms for high-dimensional optimization and big data analytics enable the methodology to extract meaningful insights and optimize processes in domains such as healthcare, finance, and engineering [20].

13.4 Multi-objective optimization and Pareto front analysis

RSM has expanded its scope to encompass multi-objective optimization problems, where multiple conflicting objectives need to be optimized simultaneously. Techniques such as Pareto front analysis and multi-criteria decision-making enable researchers to explore trade-offs and identify optimal solutions that balance competing objectives effectively [21].

13.5 Interdisciplinary applications and cross-domain collaboration

RSM continues to find new applications and opportunities for cross-domain collaboration, driven by its versatility and adaptability to diverse fields. Interdisciplinary research initiatives bring together experts from different domains, such as engineering, biology, and social sciences, to address complex problems and leverage the strengths of RSM in modeling and optimization.

14. Limitations of RSM and overcoming them

Even though response surface methodology (RSM) is a widely used statistical technique for modeling and optimizing processes with multiple variables, it does have several limitations that researchers should be aware of.

1. Assumption of polynomial model: RSM relies on the assumption that a polynomial model can adequately represent the true relationship between the response variable and the predictor variables. This assumption may not always hold true, especially for complex processes where nonlinear relationships exist.

2. Limited to small experimental regions: RSM is most effective when the experimental region is small and well-defined. When dealing with processes that have large and complex design spaces, RSM may fail to capture the entire response surface, leading to inaccurate predictions and suboptimal solutions.

3. Sensitivity to Initial Experimental Design: The choice of initial experimental design in RSM can significantly impact the accuracy of the model and the effectiveness of optimization. Poorly chosen initial designs may lead to inadequate exploration of the response surface or convergence to local optima instead of the global optimum.

4. Inability to capture interactions beyond the second order: Traditional RSM techniques are limited to capturing interactions up to the second order. For processes with higher-order interactions or non-polynomial relationships, RSM may fail to accurately model the response surface.

Overcoming the limitation: To overcome these limitations, researchers have developed several strategies and extensions to RSM:

1. Use of advanced experimental designs: Utilizing advanced experimental designs such as D-optimal or I-optimal designs can help improve the efficiency and robustness of RSM by ensuring better coverage of the design space with fewer experimental runs.

2. Incorporation of model validation techniques: Employing techniques such as cross-validation or split-sample validation can help assess the predictive accuracy of RSM models and detect potential overfitting or model misspecification.

3. Exploration of non-polynomial models: Considering alternative modeling techniques such as neural networks, kriging, or Gaussian processes can

extend the applicability of RSM to processes with nonlinear or non-polynomial response surfaces.

By addressing these limitations and leveraging appropriate techniques, researchers can enhance the effectiveness and reliability of response surface methodology in process optimization and modeling.

15. Summary and conclusion

15.1 Summary

The historical evolution of response surface methodology (RSM) marks a journey from its roots in the experimental designs of Sir Ronald Fisher to its current status as a powerful set of statistical techniques for optimizing processes and analyzing complex systems. Over the decades, RSM has witnessed significant advancements in both methodology and applications, driven by the contributions of numerous researchers across various fields.

From its inception, a methodological framework encompassing experimental design, data collection, model fitting, and optimization has characterized RSM. The introduction of Central Composite Design (CCD) by Box and Wilson in the mid-twentieth century represented a pivotal moment, enabling researchers to explore response surfaces more comprehensively with fewer experimental runs. This innovation laid the groundwork for the widespread adoption of RSM in diverse industries, including chemical engineering, manufacturing, and pharmaceuticals.

The versatility of RSM lies in its ability to handle multiple factors and their interactions, allowing researchers to uncover intricate relationships between variables and responses. This versatility has led to a myriad of applications, from optimizing chemical processes to improving product formulations and enhancing quality control. Moreover, the integration of RSM with other statistical and computational techniques has further expanded its utility, enabling researchers to tackle increasingly complex problems with confidence.

Recent developments in response surface methodology continue to propel the field forward, with advancements in experimental design, model fitting algorithms, and computational tools. The work of Myers, Montgomery, and Anderson-Cook [9] provides a comprehensive overview of RSM, emphasizing its relevance in contemporary research and industrial practice. Additionally, the review by Khuri and Mukhopadhyay [14] highlights the continued importance of RSM in the context of computational statistics, underscoring its enduring relevance in the modern era.

Looking ahead, the future of response surface methodology appears promising, with ongoing efforts aimed at refining existing techniques and exploring new avenues of application. The integration of RSM with emerging technologies such as machine learning and artificial intelligence holds great potential for enhancing predictive capabilities and accelerating the pace of discovery. Furthermore, the increasing emphasis on data-driven decision-making underscores the importance of robust statistical methods like RSM in extracting actionable insights from complex datasets.

15.2 Conclusion

The historical background of response surface methodology reflects a trajectory of innovation and discovery driven by the collective efforts of researchers spanning generations. As we continue to navigate the complexities of modern science and industry, RSM stands as a testament to the enduring power of statistical methods in shaping our understanding of the world around us. With its rich history and promising future, RSM remains a cornerstone of experimental design and process optimization, poised to make significant contributions to research and innovation in the years to come.

Author details

Hembadoon R. Asoo[1*], Joseph S. Alakali[1], Julius K. Ikya[2] and Mohammed I. Yusufu[3]

1 Centre for Food Technology and Research (CEFTER), Benue State University, Makurdi, Benue State, Nigeria

2 Department of Food Science and Technology, Joseph Sarwuan Tarka University, Makurdi, Benue State, Nigeria

3 Federal Polytechnic, Idah, Kogi State, Nigeria

*Address all correspondence to: asoo.hembadoon@gmail.com

References

[1] Fisher RA. The Design of Experiments. Edinburg, Scotland: Oliver and Boyd; 1935

[2] Box GEP, Wilson KB. On the experimental attainment of optimum conditions. Journal of the Royal Statistical Society: Series B (Methodological). 1951;**13**(1):1-45

[3] Box GEP, Behnken DW. Some new three level Designsfor the study of quantitative variables. Technometrics. 1960;**2**(4):455-475

[4] Myers RH, Montgomery DC, Anderson-Cook CM. Response Surface Methodology: Process and Product Optimization Using Designed Experiments. 3rd ed. New Jersey, USA: John Wiley & Sons; 2009

[5] Holland JH. Adaptation in Natural and Artificial Systems. Ann Arbor, USA: University of Michigan Press, University of Michigan; 1975

[6] Fisher RA. The arrangement of field experiments. Journal of the Ministry of Agriculture of Great Britain. 1926;**33**:503-513

[7] Taguchi G. Introduction to Quality Engineering: Designing Quality into Products and Processes. Tokyo, Japan: Asian Productivity Organization; 1986

[8] Montgomery DC. Design and Analysis of Experiments. 6th ed. New York, USA: John Wiley & Sons; 2005

[9] Myers RH, Montgomery DC, Anderson-Cook CM. Response Surface Methodology: Process and Product Optimization Using Designed Experiments. 4th ed. New Jersey, USA: John Wiley & Sons; 2016

[10] Box GEP. Some problems of statistics and everyday life. Journal of the American Statistical Association. 1979;**74**(365):1-4. DOI: 10.1080/01621459.1979.10481600

[11] Box GEP, Draper NR. Empirical Model-Building and Response Surfaces. New York, USA: John Wiley & Sons; 1987

[12] Box GEP, Hunter WG, Hunter JS. Statistics for Experimenters. New York, USA: John Wiley & Sons; 1978

[13] Pena D. George Box: An interview with the international journal of forecasting. International Journal of Forecasting. 2001;**17**:1-9. Available from: www.elsevier.com/locate/ijforecast

[14] Khuri AI, Mukhopadhyay S. Response surface methodology. Wiley Interdisciplinary Reviews: Computational Statistics. 2010;**2**(2):128-149

[15] Jones DR, Schonlau M, Welch WJ. Efficient global optimization of expensive black-box functions. Journal of Global Optimization. 1998;**13**:455-492. DOI: 10.1023/A:1008306431147

[16] Braga-Neto UM, Dougherty ER. Is cross-validation valid for small-sample microarray classification? Bioinformatics. 2004;**20**(3):374-380

[17] Foster I, Zhao Y, Raicu I, Lu S. Cloud computing and grid computing 360-degree compared. Grid Computing Environments Workshop, Austin, TX, USA. 2008;**2008**:1-10. DOI: 10.1109/GCE.2008.4738445

[18] Zhang Y, Wu Y. Introducing Machine Learning Models to Response Surface Methodologies. London, UK:

IntechOpen; 2021. DOI: 10.5772/intechopen.98191

[19] Brochu E, Cora VM, De Freitas N. A tutorial on Bayesian optimization of expensive cost functions, with application to active user modeling and hierarchical reinforcement learning. arXiv. 2010

[20] Van der Vaart AW, van Zanten JH. Adaptive Bayesian estimation using a Gaussian random field with inverse gamma bandwidth. Annals of Statistics. 2009;**37**(5B):2655-2675

[21] Deb K, Pratap A, Agarwal S, Meyarivan T. A fast and elitist multiobjective genetic algorithm: NSGA-II. IEEE Transactions on Evolutionary Computation. 2002;**6**(2):182-197

Chapter 2

Response Surface Techniques as an Inevitable Tool in Optimization Process

Oyebola Odunayo Olabinjo

Abstract

Response Surface Methodology (RSM) involves the construction and analysis of mathematical models to depict the relationship between input variables and the response of a system or process. This method circumvents the need for exhaustive experimentation by strategically designing a limited set of experiments while maximizing the information gathered. Experimentation and optimization are integral processes across various scientific disciplines. The utilization of Response Surface Models (RSMs) has emerged as an indispensable tool in achieving optimal experimental outcomes. The foundational understanding of RSM involves its core components, emphasizing the relationship between independent variables and their impact on a response of interest by employing statistical techniques. RSM enables researchers to comprehend the intricate behavior of systems, identify critical factors influencing the response, and subsequently optimize the process. Response surface techniques facilitates not only the improvement of processes but also the minimization of costs, reduction of waste, enhancement of product quality, facilitating efficient exploration and analysis of complex systems. Response surface analysis could be explore in all fields to generate optimal condition for all the variables in an experiment.

Keywords: optimization, response surface, independent variable, improvement process, minimization costs

1. Introduction

Optimization of experiments is of utmost importance in the field of research. It involves the identification of the optimal set of experimental conditions, which leads to the best possible result. Response Surface Methodology (RSM) is a statistical approach that is widely used for optimization of experiments in various fields, including the food industry, chemical engineering, and material science [1]. RSM is based on a mathematical model that describes the relationship between the response variable and the independent variables. This model helps to identify the optimal set of experimental conditions, which minimizes or maximizes the response variable of interest. RSM is a powerful tool for optimization of experiments, as it allows for the identification of the optimal set of experimental conditions with a minimal number of experiments. Jensen [2] highlighted the importance of RSM in product and process optimization using designed experiments. The study emphasized the efficiency of RSM in reducing the number of experiments required for optimization, which results in cost savings and faster results. Moreover, Schönbrodt et al. [3] demonstrated the use of RSM in testing similarity effects with dyadic response surface analysis. The study showed that RSM improved the efficiency of testing similarity effects. RSM has been widely used in the food industry to optimize various processes, such as extraction, drying, and fermentation [1]. For instance, Aydar [4] utilized RSM to optimize the extraction of plant materials. The study found that RSM improved the extraction efficiency, and the optimal conditions were obtained by adjusting the independent variables. Similarly, Mohammed et al. [5] used RSM to optimize the rubbercrete mixture. The study found that RSM improved the compressive strength and water absorption of the rubbercrete mixture. RSM has also been used in the optimization of chemical processes. Asfaram et al. [6] used RSM to optimize the removal of the basic dye Auramine-O from aqueous solutions. The study found that RSM improved the efficiency of the removal process, and the optimal conditions were obtained by adjusting the independent variables. Similarly, Sarabia et al. [7] discussed the use of RSM in chemical engineering, and highlighted its role in optimization of various processes. Therefore, RSM is an inevitable tool in optimization of experiments in research.

2. Principles and concepts of RSM

RSM is a statistical technique used to optimize the response of a process by exploring the relationship between the response and the input variables. The technique involves building a mathematical model that represents the relationship between the response and the input variables, which is then used to predict the response for any combination of input variables. RSM has been applied in various fields of engineering and science, including environmental engineering, manufacturing, and reliability analysis [8].

3. Types of RSM models and selection criteria

There are various types of RSM models, including central composite design (CCD), Box-Behnken design (BBD), and Taguchi design. The selection of an appropriate RSM model depends on the nature of the response variable and the number of input variables. CCD is commonly used when the response variable is continuous, and there are three or more input variables, while BBD is preferred for a smaller number of input variables. Taguchi design is used to optimize the process parameters for quality improvement [9].

3.1 Box-Behnken design

The Box-Behnken design is a type of response surface design that uses a set of three-level factorial designs with a central point. It was developed as an alternative to the more complex and time-consuming full-factorial designs. This design is particularly useful when the number of factors is high, which makes the full-factorial design impractical. The Box-Behnken design uses fewer experimental runs while maintaining a high level of accuracy in estimating the response surface. The design involves a series of levels that are evenly spaced between the high and low levels of each factor, along with a central point to estimate the curvature of the response surface.

The Box-Behnken design has several advantages over other designs. First, it has fewer experimental runs, which saves time and resources. Second, it provides a better estimation of the response surface curvature compared to other designs. Third, it allows for the assessment of the effects of each factor and their interactions, which is useful for understanding the behavior of the response variable. Finally, it is easy to implement and analyze using standard statistical software.

3.2 Central composite design

Central composite design (CCD) is a popular technique used in Response Surface Methodology (RSM) for optimization of experiments. CCD is a method for designing experiments in which the response of a system is modeled as a function of several independent variables. It is a widely used design method for fitting second-order response surface models in RSM experiments. CCD is a powerful statistical tool that is suitable for investigating non-linear relationships between experimental factors and response variables. The use of CCD in RSM has been shown to be effective in optimizing various processes, such as food processing, drug discovery, and engineering design [10, 11].

Olabinjo et al. [12] carried out the effects of oven temperature (46–74°C) and number of cycle (bath) (2–4), on the extraction yield (g/10 g) of citrus peels and antioxidant of the extract using ABTS (2,2-azinobis-(3-ethylbenzothiazoline-6-sulfonic acid)). This was investigated using statistical experimental design based on 2^2 Central Composite Design (CCD) with four axial points and five central points. The extraction employing pressurized liquid was decided to optimize only two process variables, temperature (T) and number of cycles (C). In the study, thirteen experimental treatments were assigned based on CCD with two independent variables at five levels of each variable (**Table 1**).

One of the primary advantages of CCD is its ability to model both linear and non-linear relationships between experimental parameters and response variables. CCD is also able to capture the interactions between the experimental factors, which is essential in determining the optimal conditions for a process. Ferreira et al. [13] demonstrated the efficacy of CCD in optimizing the design of frequency selective surfaces for electromagnetic applications. The authors employed CCD to model the interactions between the geometry and dimensions of the surfaces and their electromagnetic response. The results showed that CCD was able to optimize the frequency selective surfaces more effectively than other design methods.

			Levels		
Independent variables	**Axiaal ($-\alpha$)**	**Low (-1)**	**Center (0)**	**High (1)**	**Axial ($+\alpha$)**
Temperature (°C)	46	50	60	70	74
Number of cycles (C)	1.6(2*)	2	3	4	4.4(4*)

*$\alpha = \pm 1.41$ * real time in the equipment. Olabinjo et al. [12].*

Table 1.
Levels of independent variables (T and C) for the central composite design.

3.3 Taguchi design

The Taguchi design is a systematic approach to design experiments that ensure efficiency and effectiveness. It was developed by Dr. Genichi Taguchi, a Japanese engineer, in the late 1940s and early 1950s to improve the quality and reliability of products. The Taguchi design approach emphasizes the importance of choosing appropriate factors and levels to ensure that the optimization process is successful. The design approach involves conducting a series of experiments using a set of factors and levels, and then analyzing the results to determine the optimal settings for the factors. The Taguchi design approach has been used in many fields, including manufacturing, medicine, and engineering [14, 15].

The response surface model (RSM) is an effective tool for optimization of experiments using the Taguchi design approach. RSM is a mathematical model that can be used to predict the response of a system to certain inputs. The RSM can be used to optimize a system by identifying the input variables that affect the response and determining the optimal settings for these variables. The RSM can be used to develop a mathematical relationship between the input variables and the response variables. This relationship can then be used to optimize the system by predicting the optimum values for the input variables [16, 17].

The Taguchi design approach, when combined with the RSM, can lead to significant improvements in the optimization process. The Taguchi design approach can be used to identify the input variables that are most important in the optimization process, while the RSM can be used to develop a mathematical model that can predict the response of the system to these variables. The RSM can also be used to identify any interactions between the input variables. Interaction effects can significantly affect the optimization process, and it is essential to take them into account when optimizing a system [18, 19].

4. Applications of response surface methodology

Response surface methodology (RSM) has been widely used in different fields to optimize experiments. In the field of engineering, RSM has been used to optimize the design of products and processes. For instance, Panwar et al. [20] used RSM to optimize surface roughness in turning of EN 36 alloy steel. The researchers used a combination of RSM and genetic algorithm to optimize the turning process. The results showed that the combination of RSM and genetic algorithm was an effective tool for optimizing the turning process and reducing surface roughness.

In analytical chemistry, Bezerra et al. [21] used RSM to optimize the analytical parameters for the determination of metals in soil samples. The researchers used a central composite design to determine the optimum conditions for the analysis. The results showed that RSM was an effective tool for the optimization of analytical methods and could be used to reduce the number of experimental runs required for optimization.

In information systems research, Sedera and Atapattu [22] used RSM to model and optimize the relationship between the performance of information systems and the factors that influence it. The researchers used RSM to develop a response surface model that characterized the relationship between system performance and the input variables. The results showed that RSM was an effective tool for modeling and optimizing complex systems.

4.1 Advantages and limitations of response surface model

One of the most significant advantages of response surface model is that it reduces the number of experiments needed to optimize a system. RSM can identify the optimal combination of input variables using only a fraction of the experiments that would be required using the traditional one-factor-at-a-time (OFAT) method [23]. This reduction in the number of experiments not only saves time and resources but also allows researchers to explore the system more efficiently. In addition, RSM can obtain a more comprehensive understanding of the relationship between input variables and the output response, which enhances the accuracy of the optimization.

Another advantage of RSM is that it can handle multiple input variables simultaneously. Unlike OFAT, which examines one input variable at a time while holding all other variables constant, RSM can analyze the effect of multiple input variables on the output response at the same time. This feature allows researchers to identify complex interactions between input variables that would be missed by OFAT. As a result, RSM provides a more realistic optimization model that considers all the input variables in a system [24].

Furthermore, response surface model can be used to model and predict the behavior of a system. Researchers can use RSM to develop a mathematical model that describes the relationship between input variables and output response. This model can be used to predict the behavior of the system under different input variable combinations. Thus, RSM can help researchers optimize a system without conducting numerous experiments, saving time and resources while ensuring accuracy [25].

Despite the numerous advantages of response surface model, there are some limitations to its use. One limitation is that RSM assumes that the relationship between input variables and output response is continuous and smooth. If the relationship is discontinuous or has non-smooth regions, RSM may not be suitable for optimizing the system. Additionally, RSM may not be appropriate for optimizing a system that has a large number of input variables. In such cases, RSM may require a large number of experiments to optimize the system adequately, making it less practical [26].

Another limitation of response surface model is that it assumes that the input variables are independent of each other. In reality, some input variables may be correlated, so an optimization model that assumes independence may not be accurate [27]. There is also a possibility of model overfitting when there is limited data available for modeling. In such cases, the model may be overly complex, leading to poor predictions on new datasets.

5. Data analysis

The data collected determines the accuracy of the response surface model and can be carried out using various methods, such as manual collection, automated data collection, and simulation-based data collection. The data collection process should be well-designed to obtain the best results from the experimental design. In many cases, a limited number of experiments is conducted to reduce the cost and time required to obtain the data [28]. The data collected should be representative of the entire process, and the input variables should be significant factors that affect the output response. Manual collection is usually used for simple experiments that require few observations, while automated data collection is used for complex experiments that require a significant amount of data [29]. Simulation-based data collection is used to generate data by simulating the system under different parameter combinations. This method is suitable for experiments that are dangerous, costly, or time-consuming to conduct.

6. Data analysis

RSM also enables the analysis of the effects of various factors on the response variable, which can help in identifying the critical factors that affect the response variable and in improving the understanding of the system under investigation [30]. One of the main advantages of RSM is its ability to find the optimal values of the independent variables that can maximize the response variable. Li et al. [31] applied RSM to a least squares support vector machine model for annual power load forecasting, and the results showed that RSM can improve the accuracy of the model by identifying the optimal values of the input variables. Similarly, Lyu et al. [32] used RSM to optimize the aerodynamic shape of the Common Research Model wing benchmark, and the results showed that RSM can reduce the number of experimental runs required to obtain accurate results while improving the accuracy of the optimization process.

Another advantage of RSM is its ability to analyze the effects of multiple factors on the response variable. Li et al. [33] applied the d-Level Nested Logit Model to assortment and price optimization problems, and the results showed that RSM can identify the effects of different factors on the response variable, such as the price of the products and the preferences of the consumers. This information can be used to optimize the experimental design by adjusting the factors that have the most significant effects on the response variable.

7. Graphical representation of RSM result

RSM is used to establish a relationship between the response variable and the input variables. The response variable is the output of the experiment, and the input variables are the factors that influence the response variable. The response surface is a graphical representation of the relationship between the input variables and the response variable as shown in **Figure 1**. The surface plot provides a visual representation of the relationship between the input variables and the response variable. The contour plot represents the same information as the surface plot but in a two-dimensional form. The contour plot is used to determine the optimum levels of the input variables that result in the highest response variable [34, 35].

The pressurized liquid extractor was used to extract the essential oil of dried sweet orange peels of moisture content 6.5% (dry basis) using the central composite design (CCD) model of experiment. It has thirteen experiments of factorial, axial and central point as reported by Olabinjo et al. [12]; Olabinjo and Oliveira [36] as shown in **Table 1**. The PLE extractor yield result ranged from 21.1 to 49.3% with the mean of 27.7%. The positive axial temperature and negative axial static cycles had the highest yield of 49.3%. The lowest yield of 21.1% was recorded by the negative axial temperature and positive axial static cycles. Trolox was used as a reference standard for the antioxidant using ABTS, and the results were expressed as mg of trolox equivalent (mg TE) by grams of extract. In the analysis of the yield as function of process variables using response surface analysis (RSA), two optimized regions were achieved

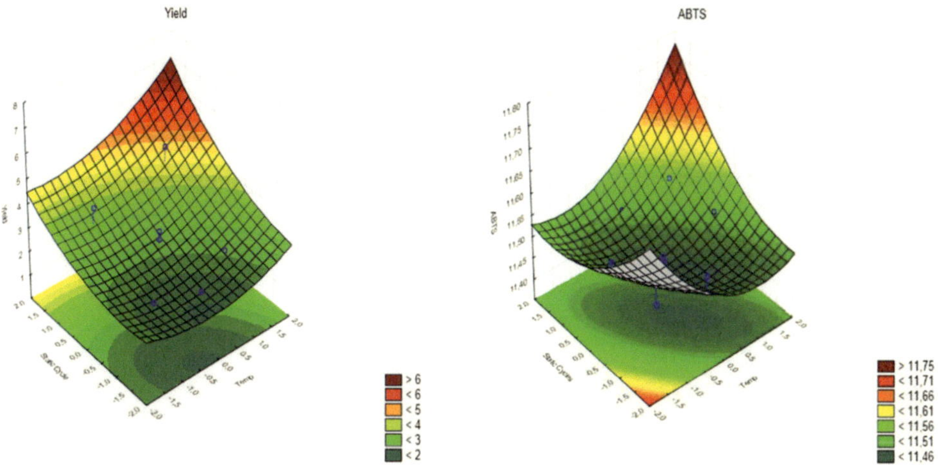

Figure 1.
Response surface of essential oil extract yield (A) and antioxidants using ABTS (B) as a function of T and SC generated by quadratic mode. Source: Olabinjo et al. [12].

(**Figure 1**) as reported by Olabinjo et al. [12]. The result from response surface analysis gave 70°C and 60 min static extraction cycle time as the optimal condition for the maximum yield, while moderate yield was reported at 46°C and 45 min of static extraction cycles. The extraction yield and antioxidant property using ABTS radical scavenging ability under the above conditions were 49.3% and 22.1%; and 11.56 and 11.53 mg trolox g^{-1} [36]. Best yield occurred when high temperature and number of cycles (C) was used and region with low temperature for a moderate number of circles, high yield was also obtained. Similar observation also occurred with the extracted antioxidant. The fact of using low temperature (T) and fewer batches or cycles (C) and thus achieve good yields and good antioxidant extracts (**Figure 1**), is characterized as an economic condition in the extraction process, with lower energy consumption and solvent. Thus, the extraction with PLE can present a great advantage over conventional methods, in which sometimes must be used at high temperatures, more solvent and take more extraction time. The low temperature of moderate yield was preferred which may prevent loss of some thermo sensitive compounds in the essential oil less solvent and reduce extraction time, power consumption is less which is economical.

Several biochemical reactions in our body generate reactive oxygen species (ROS) which are capable of damaging crucial bio-materials, if they are not effectively scavenged by cellular constituents and can lead to disease conditions. The harmful action of free radicals can be blocked by antioxidant substances, which can scavenge the free radicals and detoxify the organism. Oboh [37] reported that food materials that exhibited high antioxidant properties have the highest radical scavenging ability.

8. Conclusion

The response surface model has been shown to be an inevitable tool in the optimization of experiments. The model allows researchers to identify the optimal conditions for the variables of interest, which helps to reduce the number of experiments conducted and saves time and resources. The model can be used in various fields, including biochemical networks, aircraft design, and control systems. Response surface analysis could be explore in all fields to generate optimal condition for all the variables in an experiment.

Author details

Oyebola Odunayo Olabinjo
Department of Agricultural and Environmental Engineering, School of Engineering and Engineering Technology, Federal University of Technology, Akure, Nigeria

*Address all correspondence to: ooolabinjo@futa.edu.ng; binjo_joko@yahoo.co.uk

References

[1] Yolmeh M, Jafari S. Applications of response surface methodology in the food industry processes. Food and Bioprocess Technology. 2017;**10**:413-433

[2] Jensen WA. Response surface methodology: Process and product optimization using designed experiments 4th edition. Journal of Quality Technology. 2017;**49**:186-188

[3] Schönbrodt FD, Humberg S, Nestler S, Carlson EN. Testing similarity effects with dyadic response surface analysis. European Journal of Personality. 2018;**32**:627-641

[4] Aydar AY. Utilization of response surface methodology in optimization of extraction of plant materials. In: Silva V, editor. Statistical Approaches with Emphasis on Design of Experiments Applied to Chemical Processes. London, UK: IntechOpen Publishers; 2018. pp. 157-169

[5] Mohammed B, Khed VC, Nuruddin M. Rubbercrete mixture optimization using response surface methodology. Journal of Cleaner Production. 2018;**171**:1605-1621

[6] Asfaram A, Ghaedi M, Agarwal S, Tyagi I, Gupta V. Removal of basic dye Auramine-O by ZnS:Cu nanoparticles loaded on activated carbon: Optimization of parameters using response surface methodology with central composite design. RSC Advances. 2015;**5**:18438-18450

[7] Sarabia L, Ortiz M, Sánchez M. Response surface methodology. Comprehensive chemometrics. In: Brown S, Tauler R, Walczak B, editors. Comprehensive Chemometrics. 2nd ed. Amsterdam, The Netherlands: Elsevier; 2020. pp. 287-326

[8] Safari M, Rostami M, Alizadeh M, Alizadehbirjandi A, Nakhli SAA, Aminzadeh R. Response surface analysis of photocatalytic degradation of methyl tert-butyl ether by core/shell Fe_3O_4/ZnO nanoparticles. Journal of Environmental Health Science and Engineering. 2014;**12**:1-1

[9] An-Peng H, Xiao X, Yue R. Process parameter optimization for fused deposition modeling using response surface methodology combined with fuzzy inference system. The International Journal of Advanced Manufacturing Technology. 2014;**73**:87-100

[10] Liu R, Cheng S, Liu X, Ma L, Fan X, Luo Z. A bridging framework for model optimization and deep propagation. In: Advances in Neural Information Processing Systems 31: Annual Conference on Neural Information Processing Systems 2018, NeurIPS 2018, 3-8 December 2018, Montréal, Canada. 2018:4323-4332

[11] Joosten R, Long F, Murshudov G, Perrakis A. The PDB_REDO server for macromolecular structure model optimization. IUCRJ. 2014;**1**:213-220

[12] Olabinjo OO, Ogunlowo AS, Eliana SK, Oliveira AL. Optimization of pressurized liquid extraction of essential oil from *Citrus sinesis* peels. Agricultural Engineering International: CIGR Journal. 2020;**22**(2):255-263

[13] Ferreira D, Caldeirinha R, Cuiñas I, Fernandes T. Square loop and slot frequency selective surfaces study for equivalent circuit model optimization. IEEE Transactions on Antennas and Propagation. 2015;**63**:3947-3955

[14] Li X. Research on recommendation model optimization. Advanced Science

and Technology Letters. Vol. 79. 2014. pp. 70-74. DOI: 10.14257/astl.214.79.13

[15] Baltes P, Baltes MM. Psychological Perspectives on Successful Aging: The Model of Selective Optimization with Compensation. 1990. pp. 1-34

[16] Andersson JAE, Gillis J, Horn G, Rawlings J, Diehl M. CasADi: A software framework for nonlinear optimization and optimal control. In: Mathematical Programming Computation. 2019;**11**:1-36. DOI: 10.1007/s12532-018-0139-4

[17] Hutter F, Hoos HH, Leyton-Brown K. Sequential model-based optimization for general algorithm configuration. In: Coello CAC, editor. Learning and Intelligent Optimization. LION 2011. Lecture Notes in Computer Science. Vol. 6683. Berlin, Heidelberg: Springer; 2011. DOI: 10.1007/978-3-642-25566-3_40

[18] Kurutach T, Clavera I, Duan Y, Tamar A, Abbeel P. Model-ensemble trust-region policy optimization. ArXiv; 2018. (Cornell University)

[19] Ahn C-S, Oh S-Y. Gaussian model optimization using configuration thread control in CHMM vocabulary recognition. Journal of Digital Convergence. 2012;**10**:167-172

[20] Panwar V, Sharma D, Kumar K, Jain A, Thakar C. Experimental investigations and optimization of surface roughness in turning of en 36 alloy steel using response surface methodology and genetic algorithm. In: Materials Today: Proceedings. Vol. 46, no. 15. 2021. pp. 6474-6481

[21] Bezerra M, Santelli R, Oliveira EP, Villar LS, Escaleira LA. Response surface methodology (RSM) as a tool for optimization in analytical chemistry. Talanta: The International Journal of Pure and Applied Analytical Chemistry. 2008;**76**(5):965-977

[22] Sedera D, Atapattu M. Polynomial regression and response surface methodology: Theoretical non-linearity, tutorial and applications for information systems research. Australasian Journal of Information Systems. 2019;**23**. Research Note 1-35

[23] Ren W, Chen H. Finite element model updating in structural dynamics by using the response surface method. Engineering Structures. 2010;**32**:2455-2465

[24] Ghafari S, Aziz HA, Isa M, Zinatizadeh A. Application of response surface methodology (RSM) to optimize coagulation-flocculation treatment of leachate using poly-aluminum chloride (PAC) and alum. Journal of Hazardous Materials. 2009;**163**(2-3):650-656

[25] Vitanov V, Javaid N, Stephenson D. Application of response surface methodology for the optimisation of micro friction surfacing process. Surface & Coatings Technology. 2010;**204**:3501-3508

[26] Darnell J, Kerr I, Stark GS. Jak-STAT pathways and transcriptional activation in response to IFNs and other extracellular signaling proteins. Science. 1994;**264**(5164):1415-1421

[27] Mollon G, Dias D, Soubra A. Probabilistic analysis of circular tunnels in homogeneous soil using response surface methodology. Journal of Geotechnical and Geoenvironmental Engineering. 2009;**135**:1314-1325

[28] Najafi M, Jamali V, Schober R, Poor VH. Physics-based modeling and scalable optimization of large intelligent reflecting surfaces. IEEE Transactions on Communications. 2020;**69**:2673-2691

[29] Konecný J, McMahan HB, Ramage D, Richtárik P. Federated optimization: Distributed machine learning for on-device intelligence. arXiv.Org, abs/1610.02527. 2016

[30] Sun S, Cao Z, Zhu H, Zhao J. A survey of optimization methods from a machine learning perspective. IEEE Transactions on Cybernetics. 2019;**50**:3668-3681

[31] Li C, Li S, Liu Y. A least squares support vector machine model optimized by moth-flame optimization algorithm for annual power load forecasting. Applied Intelligence (Boston). 2016;**45**:1166-1178

[32] Lyu Z, Kenway G, Martins J. Aerodynamic shape optimization investigations of the common research model wing benchmark. AIAA Journal. 2015;**53**:968-985

[33] Li G, Rusmevichientong P, Topaloglu H. The d-level nested logit model: Assortment and price optimization problems. Operational Research. 2015;**63**:325-342

[34] Talapatra A, Boluki S, Duong T, Qian X, Dougherty E, Arróyave R. Autonomous efficient experiment design for materials discovery with Bayesian model averaging. Physical Review Materials. 2018;**2**(11). DOI: 10.1103/physrevmaterials.2.113803

[35] Eiselmayer A, Wacharamanotham C, Beaudouin-Lafon M, Mackay W. Touchstone2. HAL (Le Centre pour la Communication Scientifique Directe). 2019. DOI: 10.1145/3290605.3300447

[36] Olabinjo OO, Oliveira AL. Comparative study of extraction yield and antioxidant property of sweet orange peels (*Citrus sinesis*) essential oil. Croatian Journal of Food Science and Technology. 2020;**12**(2):184-192

[37] Oboh G. Antioxidant properties of some commonly consumed and underutilized tropical legumes. European Food Research and Technology. 2006;**224**:61-65

Section 2

Optimization Techniques and Advanced Topics

Chapter 3

Enhancing Adsorption and Desorption of Arsenic on Carbon Xerogel Nanocomposites in Aqueous Solution: Process Optimization

Sasirot Khamkure, Audberto Reyes-Rosas, Victoria Bustos-Terrones, Sofía-Esperanza Garrido-Hoyos, Prócoro Gamero-Melo and Daniella-Esperanza Pacheco-Catalán

Abstract

Arsenic, a widespread contaminant, has become a major public health concern, threatening millions globally. This study aims to develop magnetic adsorbents for easy recovery from aqueous media to remove arsenic, mitigating its environmental and health impacts. Response surface methodology (RSM) is proposed to optimize the adsorption-desorption process of As(III) and As(V) on carbon xerogel nanocomposites from aqueous solutions. A second-order polynomial model under a central composite design with a central face was employed to optimize arsenic desorption. The model parameters were estimated using least squares. Additionally, a novel linear model approach was employed to develop a second-degree polynomial model for optimizing arsenic adsorption-desorption by analyzing the effects of various factor combinations. RSM's ability to analyze the response variable over a wide range of independent variable values allows it to identify the region where the response variable reaches its optimum value.

Keywords: adsorption-desorption capacity, central composite design, magnetite nanoparticle, regeneration, response surface methodology

1. Introduction

1.1 Arsenic contamination and its health limitations

Trivalent arsenic (As(III), arsenites) and pentavalent arsenic (As(V), arsenates), are well-established carcinogens and significant environmental and public health hazards. Notably, arsenic has emerged as a leading global chemical contaminant due to its natural presence at elevated levels in groundwater, particularly in arid northern and central Mexico, where weathering of silica volcanic rock results in its migration into the water [1]. The presence of arsenic in the environment has significantly emerged as a major public health problem in surpassing the World Health Organization's maximum permissible limit of 10.0 μg/L, it poses a potential risk of adverse health effects, including cancer, peripheral neuropathy, arsenicosis, and cardiovascular diseases. This threat is further amplified by the element's high solubility in water [2, 3]. In Mexico, the maximum permissible limit for arsenic in drinking water is also set at 10.0 μg/L, as outlined in NOM-127-SSA1-2021.

1.2 Addressing arsenic with adsorption

Adsorption using porous materials is a widely employed technique for treating drinking water and wastewater [4]. This method allows for the selective removal of specific contaminants by choosing an appropriate adsorbent. To enhance the economic and environmental sustainability of this process, studying the desorption and reuse of adsorbents is crucial [5, 6]. In this case, composite gels with magnetite nanoparticles were prepared to have electrostatic attraction of nanoparticles of iron oxide, high surface area, easy to recover from the aqueous medium, and reusable of magnetic polymers [7, 8]. One of the most important factors in designing an adsorption process is the quantity and types of pollutants (adsorbents) present in the water source. These pollutants can interact competitively with each other, or simply reduce the adsorbent's functionality through early saturation or synergies within the aquatic system. This decreased functionality further complicates the availability of clean drinking water, especially when dealing with contaminants like inorganic arsenic compounds present in water bodies.

1.3 Application of response surface methodology

Knowledge and understanding of adsorption process design and optimization are essential. Several factors must be considered for effective adsorption, such as pH, synthesis method, initial concentration, particle size, competing ions, and contact medium [9, 10]. Experimental design is a valuable tool for optimizing and predicting the interactive effects of adsorption processes [10, 11]. Response surface methodology (RSM) is a mathematical and statistical method widely used to evaluate multiple factors and their interactions with various response variables [12]. RSM has been effectively applied in various studies to optimize the adsorption process. RSM has been used to examine the interactive effects and predict the model building and optimization of arsenic adsorption by activated carbon derived from food industry waste [9], biodegradable biopolymer chitosan [13], and natural zeolite with magnesium oxide [14]. RSM was used to optimize the adsorption of heavy metals, nickel, and cadmium, respectively [15, 16], from aqueous solutions. RSM was also utilized to optimize the ultrasound-assisted adsorption of anionic and cationic dyes, demonstrating the versatility of RSM in different adsorption processes [17]. These studies collectively highlight the effectiveness of RSM in optimizing the adsorption process for various contaminants.

1.4 Experimental design with RSM

In this study, the adsorption and desorption of arsenic on carbon xerogel nanocomposites propose establishing and carrying out RSM. RSM employs a series of mathematical and statistical techniques to model and analyze problems where the goal is to obtain the best-fitting mathematical model that relates a dependent response variable (often denoted as "y") to various independent variables ("x_i").

1.5 Advantages of RSM over ordinary factorial designs

Unlike traditional factorial designs (e.g., 2^k), RSM allows us to identify the optimal operating conditions for a process. This refers to the specific combination of "x_i" values that maximize or minimize the response variable. This achievement is possible through the utilization of more advanced mathematical techniques. While a factorial design can identify a "winning" combination or treatment within the tested values, it cannot necessarily be extrapolated to untested values.

RSM offers several advantages over ordinary factorial designs. It minimizes the number of experiments required for a specific number of factors and levels, making it more efficient [18]. Additionally, RSM can be used to construct cost-efficient response surface designs for experiments with both qualitative and quantitative factors [19]. These advantages collectively make RSM a powerful tool for experimental design and optimization.

1.6 Strength of RSM

The key strength of RSM lies in its ability to interpolate between different values of the independent variables. This means we can not only identify the optimal combination from the tested data but also predict the optimal values for untested combinations. This prediction is done by constructing 3D surface graphics or 2D contour plots of the response variable.

2. The experiment design for As(III) and As(V) adsorption on carbon xerogel nanocomposites

This study aims to develop magnetic adsorbents that remove both As(III) and As (V) efficiently from groundwater. Carbon xerogel nanocomposites were synthesized via sol-gel polymerization of a resorcinol-formaldehyde composite incorporating magnetite nanoparticles [7]. The process utilized sonication-assisted direct methods, followed by carbonization and H_2O_2-induced surface modification. This study evaluated the effects of dosage, solution pH, and initial concentration of arsenic on the removal process using an experimental design. RSM was used to optimize the adsorption process of arsenic on magnetic carbon xerogel nanocomposites from aqueous solutions using a central composite design (CCD) 2^3 with a central face.

2.1 Experimental design and optimization method

The magnetic carbon xerogel nanocomposite was selected for study using a CCD and evaluated for arsenic removal using RSM. RSM allows the interpolation of various independent variable values and identifies optimal combinations for the response variable by constructing 3D surface graphs and projecting their contours onto a 2D surface.

The RSM application is proposed through a second-order polynomial model under a CCD scheme with a central face. The least squares method estimates parameters in this model. For each independent variable combination ("x_i"), 3D graphs and 2D contour plots will be obtained, along with various statistical parameters evaluating their fitting and reliability, such as: R-squared (R^2), adjusted R^2, F test statistic, lack of fit test, and locating the stationary point (maximum response point).

This method can be implemented using statistical software R v4.2. **Table 1** summarizes the evaluation method, which consists of one response variable (As removal efficiency) and three independent variables (adsorbent dose, initial As concentration, and solution pH) at three levels (low, medium, and high).

2.2 Arsenic batch adsorption experiment

For the batch adsorption experiment, sodium arsenite ($NaAsO_2$, Sigma-Aldrich) and sodium arsenate dibasic heptahydrate ($HAsNa_2O_4{\cdot}7H_2O$, Sigma-Aldrich) were used to prepare As(III) and As(V) stock solutions at various desired concentrations. These solutions were prepared by diluting with ultrapure (type 1) water. The pH of the solution was adjusted using 0.1 M HCl and 0.1 M NaOH solutions and then measured with a pH meter (Orion Star A211, Thermo Scientific, USA).

Following the conditions outlined in **Table 1**, all experiments were conducted in 50 mL centrifuge tubes at 150 rpm with a room temperature range of 25°C to 28°C. All the results obtained from the experiment were evaluated for removal efficiency and adsorption capacity.

Equations:

The removal efficiency was evaluated using the following equation (Eq. (1)):

$$\%\text{removal} = (C_0 - C_e)/C_0 \times 100 \quad (1)$$

where:

- C_0: initial concentration of arsenic (mg/L)
- C_e: final concentration of arsenic (mg/L)

The adsorption capacity (q_e) was calculated using formula (Eq. (2)):

$$q_e = (C_0 - C_e)/m \times V \quad (2)$$

where:

- m: Mass of adsorbent (g)
- V: Solution volume (L)

Factors	Coding factors	Low (−1)	Center (0)	High (1)	Units
Solution pH	x_1	3.0	5.0	7.0	pH value
Adsorbent dose	x_2	0.5	2.5	4.5	g/L
As concentration	x_3	0.05	0.15	0.25	mg/L

Table 1.
Coded factors and levels for As(V) and As(III) adsorption experiment.

2.3 Determination of As(V) and As(III) concentrations in water samples

After the adsorption experiment, each solution was filtered using a vacuum pump and 0.45 μm cellulose nitrate membrane filters (Whatman) to separate the solid from the supernatant. The concentrations of As(V) and As(III) were determined using a fast sequential flame atomic absorption spectrometer (SpectrAA-220, Varian, Australia) at a wavelength of 193.7 nm [20].

2.4 Models fitted and their corresponding optimal parameters

The evaluation of the removal of arsenic using magnetic carbon xerogel nanocomposite was carried out by executing an RSM. The response variable was arsenic removal efficiency, and the independent variables were adsorbent dose, initial arsenic concentration, and solution pH. Each variable had three levels: Low, medium, and high. The experimental method and data analysis were implemented using statistical software R v4.2

Eqs. (3) and (4) present the most significant parameters identified using RSM for As(V) and As(III) removal efficiency, respectively. Categorical variables were coded according to **Table 1**.

$$\text{As(V) removal [\%]} = 28.6177 - 23.4317x_1 + 10.8115x_2 - 11.9858x_1x_2 + 26.0798x_1^2 - 9.5217x_3^2 \quad (3)$$

$$\text{As(III) removal [\%]} = 51.96523 - 7.42122x_1 + 13.36756x_2x_3 - 18.22614x_1^2 + 15.53519x_3^2 \quad (4)$$

Figures 1 and **2** show the contour plots of the removal efficiency of As(V) and As (III) with two different factors, respectively. The x and y axes show the two variable input parameters (pH, dose, or initial concentration) that maximize arsenic removal. The most efficient As(V) removal was achieved for the significant quantitative factors

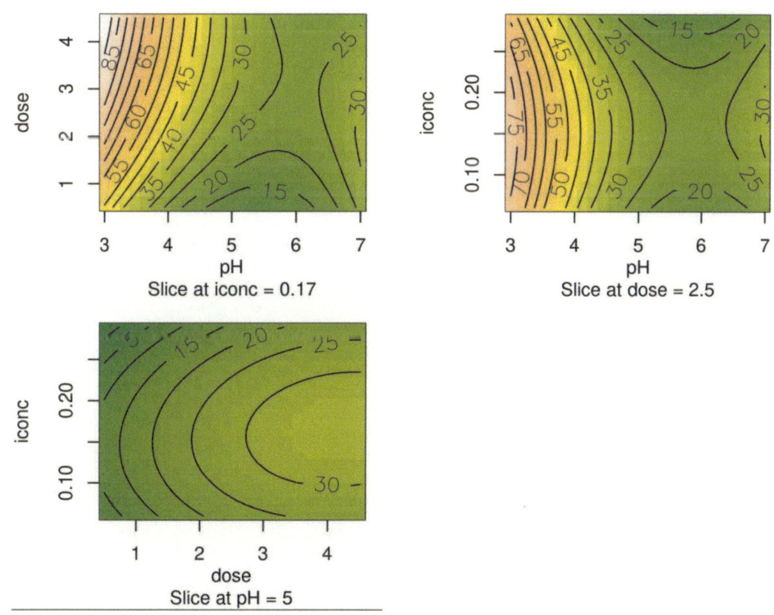

Figure 1.
The 2D representation visualizes the optimized results obtained by applying the RSM to optimize pH, initial concentration, and dose for the removal efficiency of As(V).

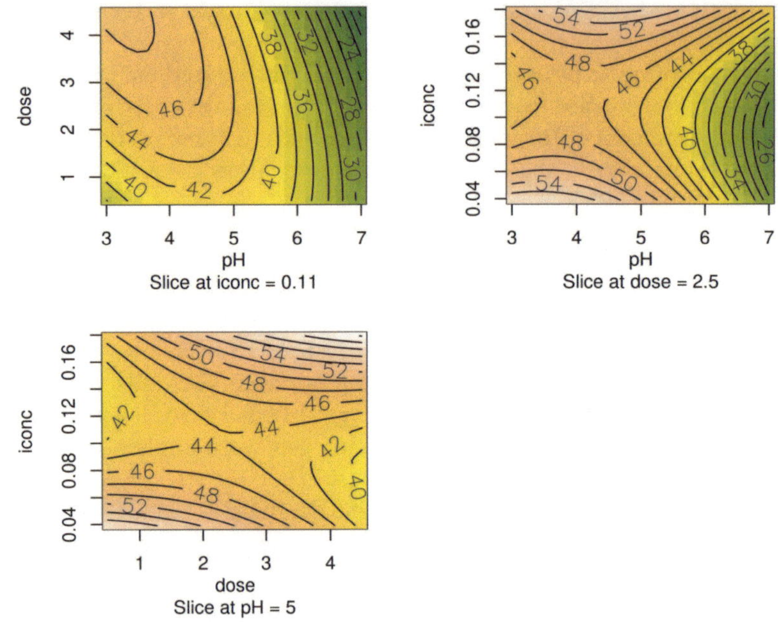

Figure 2.
The 2D representation visualizes the optimized results obtained by applying the RSM to optimize pH, initial concentration, and dose for the removal efficiency of As(III).

pH and adsorbent dose (**Figure 1**). **Figure 2** shows the contour plot of the As(III) removal efficiency. The highest As(III) removal efficiency was obtained for the significant quantitative factors of adsorbent dose and initial concentration.

Figure 3 presents the adjusted model based on significant parameters, resulting from the RSM. It visualizes the removal efficiency of As(V) as a function of pH vs dose.

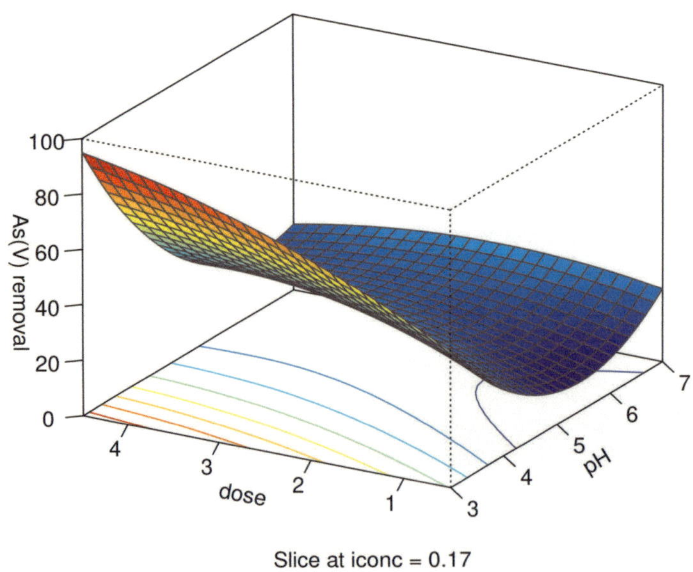

Figure 3.
The 3D surface estimation plots of As(V) removal efficiency as a function of pH vs. dose.

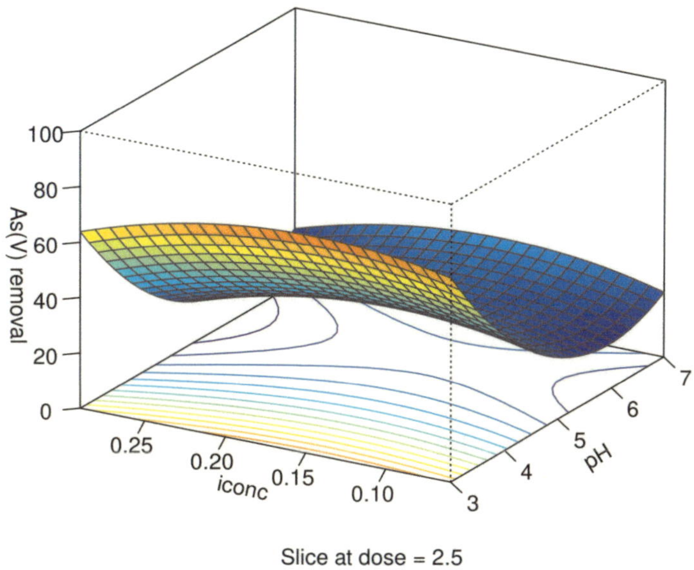

Figure 4.
The 3D surface estimation plots of As(V) removal efficiency as a pH and initial concentration.

Figure 4 presents the adjusted model resulting from the RSM, visualizing the removal efficiency of As(V) as a function of pH and initial concentration. The maximum As(V) removal efficiency was obtained at a low pH and a high adsorbent dose. Studies on the adsorption of various compounds by carbon xerogel have found similar results, suggesting that both pH and the dose of the xerogel play a significant role [21, 22].

The optimal values were determined by calculating the estimated marginal means for factor combinations in a linear model using the "emmeans" package in R [23]. The optimum values for pH, adsorbent dose, and initial concentration for the removal of As(V) were 3.0, 4.5 g/L, and 0.15 mg/L, respectively, with a removal efficiency of 95.03 $\pm$ 5.98%, which was achieved within 3 h in aqueous solutions. The optimum values for pH, adsorbent dose, and initial concentration for the removal of As(III) were 4.0, 4.5 g/L, and 0.18 mg/L, respectively, with a removal efficiency of 65.04 $\pm$10.32% (**Table 1**).

3. The experiment design for arsenic desorption on carbon xerogel nanocomposites

In this study, the desorption capacity, and the possibility of regeneration of the carbon xerogel nanocomposites were evaluated in the arsenate adsorption process from an aqueous solution. The optimal condition for the arsenate desorption process was studied using a RSM, varying the concentration of desorbing agents, adsorbent doses, and the agitation at which the magnetic gels were prepared.

The use of RSM is proposed by using a 16-run CCD with a central point and fitting a second-order polynomial model. The parameters of the model are estimated using the least squares method.

The 3D graphs and 2D contour plots are obtained for combinations of the independent variables "x_i". Statistical parameters like R-squared, adjusted R-squared,

F-statistic, lack of fit test, and the location of the stationary point (the point on the response surface where the percentage removal of arsenic is maximized) are used to evaluate the fit and reliability of the model.

This method can be implemented using the statistical software R (version 4.2). The setup is consisted of:

One response variable: Percentage removal of arsenic.

Three independent variables at three levels (low, medium, and high values): The concentration of desorbing agents, adsorbent doses, and the agitation.

A central face: Treatment with mean values of the three independent variables.

The experimental data are composed of 14 different combinations of the independent variables plus two additional "central face" treatments in two replicates for a total of 32 measurements.

3.1 The optimal condition for the arsenic desorption process using RSM

The optimization study of the desorption of arsenic was carried out by using a RSM. The experimental method and data analysis were implemented using the software R. The experimental design of CCD with three factors and two center points was applied in this study with the varying concentration of desorbing agent (M), spent adsorbent dose (g/L), and orbital shaker speed (rpm) as shown in **Table 2**.

3.2 Analysis of results implementing the RSM for arsenic desorption

For the experimental design of the RSM configuration for optimizing arsenic desorption using KOH on carbon xerogel nanocomposites [24], results of a second-degree polynomial model are shown in **Tables 3** and **4**. The variables encoded were x_1 (KOH concentration), x_2 (Agitation speed), and x_3 (Adsorbent dose).

The variable x_3 (dose) has a significantly stronger effect (indicated by more asterisks) on the response variable As desorption.

The resulting second-degree model is:

$$\begin{aligned} \text{As desorption} = {} & 49.07 + 2.4x_1 + 3.78x_2 + 29.37x_3 - 2.54x_1x_2 + 3.84x_1x_3 \\ & + 4.98x_2x_3 - 8.9x_1^2 + 11.04x_2^2 - 3.17x_3^2 \end{aligned} \quad (5)$$

With the coded variables x_1(Conc), x_2(Speed), and x_3(Dose).

R-squared: 0.9237, Adjusted R-squared: 0.8092

F-statistic: 8.067, 9 DF, 6 DF, p-value: 0.009712

The p-value of the model is significant (p-value < 0.05).

The coefficient of lack of fit is not significant (Lack of fit > 0.05), which means the fitting level for the model is acceptable.

Factors of arsenic desorption	Coding factors	Low (−1)	Center (0)	High (1)
Concentration of KOH solution (M)	x_1	0.5	1	1.5
Orbital shaker speed (rpm)	x_2	80	120	160
Spent adsorbent dose (g/L)	x_3	0.4	1.2	2

Table 2.
Configuration of variables for implementation of the RSM method to optimize arsenic desorption on carbon xerogel nanocomposites.

Variable	Estimate	Std. error	t value	Pr(>|t|)
(Intercept)	49.0661	5.4488	9.0050	0.0001049***
x_1	2.3998	3.6395	0.6594	0.5341302
x_2	3.7816	3.6395	1.0391	0.3388262
x_3	29.3671	3.6395	8.0690	0.0001940***
$x_1{:}x_2$	−2.5448	4.0691	−0.6254	0.5547302
$x_1{:}x_3$	3.8370	4.0691	0.9430	0.3820983
$x_2{:}x_3$	4.9842	4.0691	1.2249	0.2665189
x_1^2	−8.8971	7.0882	−1.2552	0.2560800
x_2^2	11.0396	7.0882	1.5574	0.1703725
x_3^2	−3.1746	7.0882	−0.4479	0.6699693

*Significant Codes: 0 '***' 0.001 '**' 0.01 '*' 0.05 '.' 0.1 ' ' 1*

Table 3.
Statistical analysis of the second-degree polynomial model generated using the RSM method.

Df	Sum	Sq mean	Sq F value	Pr(>F)	
rsm:FO (x_1, x_2, x_3)	3	8824.9	2941.62	22.2077	0.001195
TWI (x_1, x_2, x_3)	3	368.3	122.78	0.9269	0.483106
PQ (x_1, x_2, x_3)	3	424.1	141.36	1.0672	0.430404
Residuals	6	794.8	132.46		
Lack of fit	5	677.0	135.39	1.1495	0.606208

Table 4.
RSM analysis of the second-degree polynomial model's statistics.

Figure 5 demonstrates the behavior of arsenic desorption with respect to the variables considered, using the stationary points in their original units, namely, concentration (conc) of 1.64, speed of 77.6, and dose of 4.85. It can be observed that the agitation speed and the concentration of the desorbing agent increase, and the arsenic desorption of the material improves.

Figure 6 depicts a 3D surface plot of arsenic desorption percentage as a function of both desorbing agent concentration (KOH) and adsorbent dose. The plot is generated using the RSM method and utilizes an agitation speed of 77.6 rpm, determined as the stationary point through analysis. As visually indicated by the increased number of asterisks, the adsorbent dose (variable x_3) exerts a significant influence on the response variable, arsenic desorption. Furthermore, the model demonstrates statistical significance with a p-value less than 0.05.

Table 5 presents combinations of values that maximize the As desorption parameters within specified ranges: KOH solution concentration (conc) = 0.5–3.0 M, orbital shaker speed (speed) = 120–180 rpm, and dose of spent adsorbent (dose) = 0.4–5.0 g/L. The highest As desorption values are based on the best parameter combinations.

Desorption of arsenic from spent carbon xerogel nanocomposites predicted by the second-degree model estimated with RSM, visualized as a 3D plot as shown in **Figure 7**. Dose and concentration are the model variables, while speed is fixed at

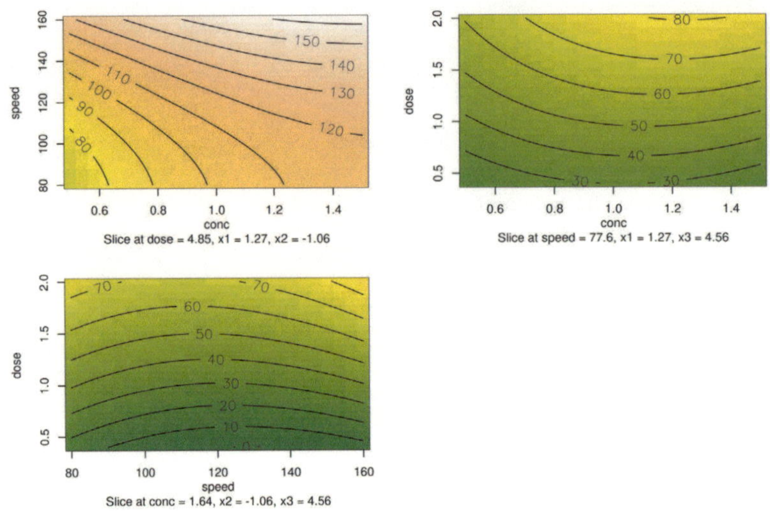

Figure 5.
The 2D representation (contour lines) of As desorption behavior as a function of concentration, speed, and dose, using the stationary points: Concentration = 1.64, speed = 77.6, and dose = 4.85, found by the RSM method.

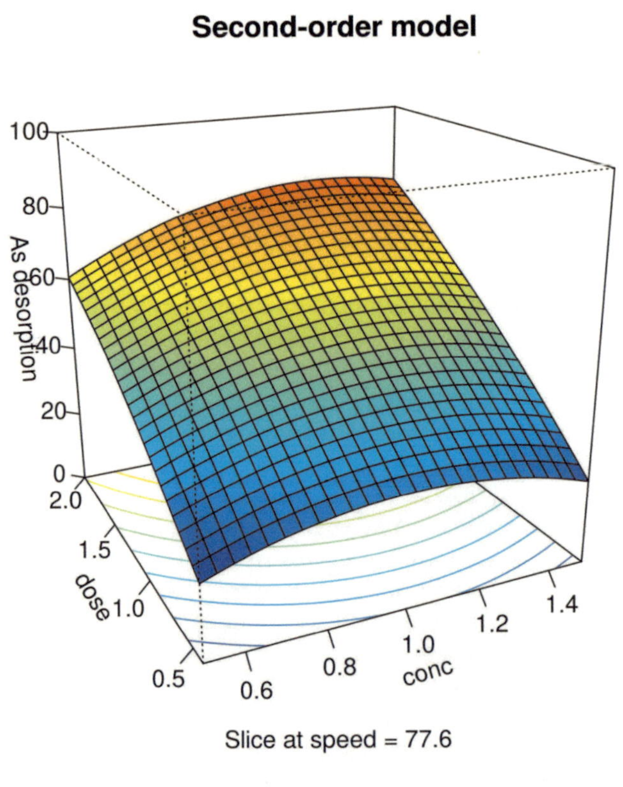

Figure 6.
The 3D representation of As desorption behavior as a function of concentration (conc) versus dose, with speed = 77.6 fixed as the stationary point estimated by the RSM method.

conc	speed	dose	As_des_avg	SE	df	lower.CL	upper.CL
1.05	160	2	95.3442	9.7559	6	71.4725	119.216
1.25	160	2	94.6857	9.4133	6	71.6521	117.7193
0.85	160	2	93.1557	9.6096	6	69.6417	116.6696
1.05	160	1.9	91.7464	8.9601	6	69.8219	113.6709
1.05	150	2.1	91.7295	9.8582	6	67.6073	115.8516
1.25	150	2.1	91.5173	9.5527	6	68.1427	114.8918
1.45	160	2	91.1801	9.7877	6	67.2306	115.1297
1.25	160	1.9	90.896	8.6362	6	69.7641	112.028
0.85	160	1.9	89.7497	8.817	6	68.1752	111.3242
0.85	150	2.1	89.0946	9.7244	6	65.2998	112.8893
1.05	150	2	88.3866	8.4758	6	67.647	109.1262
0.65	160	2	88.12	9.3719	6	65.1878	111.0522
1.05	160	1.8	88.0494	8.4758	6	67.3098	108.7889
1.25	150	2	87.9826	8.1787	6	67.9699	107.9952
1.45	160	1.9	87.1986	9.1539	6	64.7997	109.5974
1.25	160	1.8	87.0072	8.1787	6	66.9945	107.0198
0.85	160	1.8	86.2445	8.3395	6	65.8385	106.6505
1.25	140	2.1	86.0383	9.4231	6	62.9808	109.0959
1.05	140	2.1	85.996	9.6731	6	62.3267	109.6653
0.85	150	2	85.9435	8.3395	6	65.5376	106.3495

Conc = Concentration of initial; speed = Speed of agitation; dose = Dose of adsorbent; As_des_avg = Average of arsenic desorption; SE = Standard error; df = degrees of freedom; lower.CL = lower confidence level; upper.CL = upper confidence level.

Table 5.
Combinations of values maximizing the As desorption parameters subject to the restriction that the variables range between concentration, speed, and dose.

160 rpm (as specified in **Table 2**) and was used to maximize As desorption. Similar findings show that the removal efficiency of arsenic increased with higher adsorbent dosages, indicating a potential for strong adsorption [25, 26].

Limitations of the implemented methodology are related to the probable need to perform different modeling when using the RSM method to find the best combinations of predictor variables that optimize the response variable. Otherwise, the model and the obtained response will only be valid for the range of values considered in the modeling with the risk of obtaining a local maximum instead of a global maximum value that considers a wider range of values of the predictor variables.

Future research could focus on comparing the results of the RSM method with models created from machine learning techniques. These techniques generally require a larger amount of data for their construction but could have better adaptation to nonlinear behaviors and compensate for those situations in which the RSM method shows difficulty in fitting.

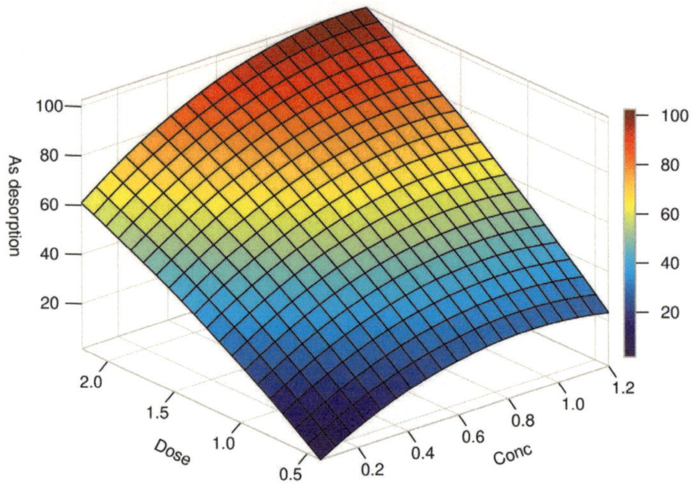

Figure 7.
Implementation of the second-degree model estimated using the RSM method based on the dose vs conc variables, with a constant value of speed = 160 rpm.

4. Conclusions

This study presents a significant advancement in arsenic removal technology. Carbon xerogel nanocomposites loaded with magnetite nanoparticles were synthesized using a streamlined process involving direct sonication during sol-gel polycondensation. Subsequent carbonization and hydrogen peroxide surface modification further enhanced the material's adsorption capacity.

Employing RSM proved to be a powerful tool. It revealed not only the critical factors influencing arsenic adsorption (solution pH and nanocomposite dose for both As(III) and As(V)) but also established a predictive model for optimizing arsenic capture under various conditions. This paves the way for efficient and targeted arsenic removal strategies.

The impact of RSM extended beyond adsorption. It facilitated the optimization of the desorption process, identifying the adsorbent dose as the key factor governing arsenic release. Moreover, a novel RSM-based model was developed to predict the effects of combined factors on desorption. This innovative approach offers valuable insights for future research and practical applications.

The next phase of this research will prioritize evaluating the efficacy of these nanocomposites under environmentally relevant conditions to assess their arsenic removal efficiency in a more complex matrix. The reusability of the As-loaded nanocomposite will be investigated through multiple adsorption-desorption cycles. Potential changes in performance will be analyzed, and regeneration methods explored to maintain long-term efficiency. This comprehensive investigation will ensure the technology's effectiveness and sustainability for practical arsenic remediation.

Acknowledgements

Khamkure S. acknowledges the 'Investigadoras e Investigadores por México CONACYT' program (Project No. CIR/0069/2022). This work was partially supported by Universidad Autónoma Agraria Antonio Narro (Grant No. 38111-425401001-2320). The authors would also like to thank Manuel Sanchez-Zarza, Socorro García-Guillermo, and José Martín Baas-López for their contributions with technical equipment and analysis.

Conflict of interest

The authors declare no conflict of interest.

Thanks

The authors thank María-Fernanda Cabello-Lugo and Arael Torrecilla-Valle for their helping and supporting in the laboratory.

Author details

Sasirot Khamkure[1*], Audberto Reyes-Rosas[2], Victoria Bustos-Terrones[3], Sofía-Esperanza Garrido-Hoyos[4], Prócoro Gamero-Melo[5] and Daniella-Esperanza Pacheco-Catalán[6]

1 Irrigation and Drainage Department, CONAHCYT-Antonio Narro Agrarian Autonomous University, Saltillo, Mexico

2 Department of Bioscience and Agrotechnology, Research Center of Applied Chemistry, Saltillo, Mexico

3 Engineering in Environmental Technology, Polytechnic University of the State of Morelos, Jiutepec, Mexico

4 Postgraduate Department, Mexican Institute of Water Technology, Jiutepec, Mexico

5 Sustainability of Natural Resources and Energy, CINVESTAV Saltillo, Ramos Arizpe, Mexico

6 Renewable Energy Unit, Yucatan Scientific Research Center, Merida, Mexico

*Address all correspondence to: skhamkure@conahcyt.mx

References

[1] Alarcón-Herrera MT, Martin-Alarcon DA, Gutiérrez M, Reynoso-Cuevas L, Martín-Domínguez A, Olmos-Márquez MA, et al. Co-occurrence, possible origin, and health-risk assessment of arsenic and fluoride in drinking water sources in Mexico: Geographical data visualization. Science of the Total Environment. 2020;**698**: 134168. DOI: 10.1016/j.scitotenv.2019.134168

[2] Wei B, Yu J, Kong C, Li H, Yang L, Guo Z, et al. An investigation of the health effects caused by exposure to arsenic from drinking water and coal combustion: Arsenic exposure and metabolism. Environmental Science and Pollution Research. 2017;**24**(33): 25947-25954

[3] Valskys V, Hassan HR, Wołkowicz S, Satkūnas J, Kibirkštis G, Ignatavičius G. A review on detection techniques, health hazards and human health risk assessment of arsenic pollution in soil and groundwater. Minerals. 2022;**12**(10):1-30

[4] Jose Alguacil F, Ignacio Robla J. Arsenic and biosorption. In: Arsenic in the Environment – Sources, Impacts and Remedies. [Internet]. IntechOpen; 2023. Available from: https://www.intechopen.com/chapters/1135908

[5] Bibi S, Kamran MA, Sultana J, Farooqi A. Occurrence and methods to remove arsenic and fluoride contamination in water. Environmental Chemistry Letters. 2017;**15**:125-149

[6] Indah S, Helard D, Binuwara A. Studies on desorption and regeneration of natural pumice for iron removal from aqueous solution. Water Science and Technology. 2017;**2017**(2): 509-515

[7] Khamkure S, Bustos-Terrones V, Benitez-Avila NJ, Cabello-Lugo MF, Gamero-Melo P, Garrido-Hoyos SE, et al. Effect of $Fe3O_4$ nanoparticles on magnetic xerogel composites for enhanced removal of fluoride and arsenic from aqueous solution. Water Science and Engineering. 2022;**15**(4): 305-317

[8] Khamkure S, Gamero-Melo P, Garrido-Hoyos SE, Reyes-Rosas A, Pacheco-Catalán DE, López-Martínez AM. The development of Fe_3O_4-monolithic resorcinol-formaldehyde carbon xerogels using ultrasonic-assisted synthesis for arsenic removal of drinking water. Gels. 2023;**9**(8):1-29

[9] Bakkal Gula C, Bilgin Simsek E, Duranoglu D, Beker U. An experimental design approach for modeling As(V) adsorption from aqueous solution by activated carbon. Water Science and Technology. 2015;**71**(2):203-210

[10] Mtaallah S, Marzouk I, Hamrouni B. Factorial experimental design applied to adsorption of cadmium on activated alumina. Journal of Water Reuse and Desalination. 2018;**8**(1):76-85

[11] Özbay N, Yargiç AŞ, Yarbay-Şahin RZ, Önal E. Full factorial experimental design analysis of reactive dye removal by carbon adsorption. Journal of Chemistry. 2013;**2013**:1-13

[12] Aydar AY. Utilization of response surface methodology in optimization of extraction of plant materials. In: Silva V, editor. Statistical Approaches With Emphasis on Design of Experiments Applied to Chemical Processes. Rijeka: IntechOpen; 2018. p. Ch. 10. DOI: 10.5772/intechopen.73690

[13] Perez Mora B, Bellú S, Mangiameli MF, Frascaroli MI, González JC. Response surface methodology and optimization of arsenic continuous sorption process from contaminated water using chitosan. Journal of Water Process Engineering. 2019;**32**(July):100913. DOI: 10.1016/j.jwpe.2019.100913

[14] Mejía-Zamudio F, Valenzuela-García JL, Aguayo-Salinas S, Meza-Figueroa D. Adsorción de arsénico en zeolita natural pretratada con óxidos de magnesio. Revista Internacional de Contaminacion Ambiental. 2009;**25**(4): 217-227

[15] Turki N, Boujelben N, Bakari Z, Bouzid J. Use of response surface methodology for optimization of nickel adsorption in an aqueous solution by clay. American Journal of Environmental Sciences. 2021;**17**(5):92-100

[16] Dizaj Khalili A, Ghaemi A. Utilization of response surface methodology, kinetic and thermodynamic studies on cadmium adsorption from aqueous solution by steel slag. Journal of the Iranian Chemical Society. 2021;**18**(11):3031-3045

[17] Dastkhoon M, Ghaedi M, Asfaram A, Goudarzi A, Mohammadi SM, Wang S. Improved adsorption performance of nanostructured composite by ultrasonic wave: Optimization through response surface methodology, isotherm and kinetic studies. Ultrasonics Sonochemistry. 2017;**37**:94-105

[18] Chelladurai SJS, Murugan K, Ray AP, Upadhyaya M, Narasimharaj V, Gnanasekaran S. Optimization of process parameters using response surface methodology: A review. Materials Today: Proceedings. 2020;**37**(2):1301-1304

[19] Divecha J, Tarapara B. Small, balanced, efficient, optimal, and near rotatable response surface designs for factorial experiments asymmetrical in some quantitative, qualitative factors. Quality Engineering. 2017;**29**(2): 196-210

[20] Quevedo O, Luna B. Determinación de As (III) y As (V) en aguas naturales por generación de hidruro con detección por espectrometría de absorción atómica. Revista CENIC Ciencias Químicas. 2003;**34**(3):133-147

[21] Sherlala AIA, Raman AAA, Bello MM, Buthiyappan A. Adsorption of arsenic using chitosan magnetic graphene oxide nanocomposite. Journal of Environmental Management. 2019;**246**:547-556

[22] Arsić TM, Kalijadis A, Matović B, Stoiljković M, Pantić J, Jovanović J, et al. Arsenic(III) adsorption from aqueous solutions on novel carbon cryogel/ceria nanocomposite. Processing and Application of Ceramics. 2016;**10**(1): 17-23. Available from: http://www.doiserbia.nb.rs/Article.aspx?ID=1820-61311601017M

[23] Lenth R. Emmeans: Estimated Marginal Means. Aka Least-Squares Means. 2023. Available from: https://cran.r-project.org/package=emmeans

[24] Hassan AF, Abdel-Mohsen AM, Elhadidy H. Adsorption of arsenic by activated carbon, calcium alginate and their composite beads. International Journal of Biological Macromolecules. 2014;**68**:125-130

[25] Yao S, Liu Z, Shi Z. Arsenic removal from aqueous solutions by adsorption onto iron oxide/activated carbon magnetic composite. Journal of Environmental Health Science and Engineering. 2014;**12**(1):1-8

[26] Hua J. Adsorption of low-concentration arsenic from water by co-modified bentonite with manganese oxides and poly (dimethyldiallylammonium chloride). Journal of Environmental Chemical Engineering. 2018;**6**(1):175-186

Chapter 4

Formulae for Slopes and Curvatures of Two Curves in Quadratic Polynomial Regression Equations with Response Surface Analysis

Shaoping Qiu and Kaizhe (Kaiser) Qiu

Abstract

Response Surface Methodology (RSM) constitutes a suite of statistical and mathematical techniques employed for the development, enhancement, and optimization of processes with the aim of achieving the maximum (or minimum) value of a response variable. In the realm of organizational research, scholars have embraced RSM, utilizing quadratic regression equations illustrated on a three-dimensional surface to investigate congruence phenomena such as fit, match, similarity, or agreement. To date, RSM has found extensive application in exploring nuanced relationships among combinations of two predictor variables and an outcome variable within organizational studies. This paper takes a novel approach by incorporating directional derivatives and the rotation of axes from multivariable calculus. It formulates a set of equations to calculate the slopes and curvatures of two curves in quadratic polynomial regression equations through response surface analysis.

Keywords: response surface methodology, quadratic polynomial regression, formula, curve, slope

1. Introduction

Response surface methodology (RSM) serves as an expansive toolkit, encompassing statistical and mathematical techniques with the overarching goal of developing, improving, and optimizing processes to achieve the maximum or minimum value of a response [1]. This empirical model employs mathematical and statistical approaches to establish relationships between a response and various input variables that may influence it. In the field of statistics, RSM has proven invaluable for scrutinizing the intricate relationships between predictor and response variables.

Originally conceptualized by Box and Wilson [2], RSM was initially applied to explore the correlations between the yield of a chemical process and a set of input variables presumed to impact the yield [3]. The fundamental principle of RSM involves utilizing a second-degree polynomial model to achieve an optimal outcome through a series of experiments. Its simplicity in estimation and application makes this statistical technique particularly valuable for maximizing the results of a specific substance through the optimization of operational factors *via* well-designed and analyzed experiments [3].

In the context of organizational research, significant credit is attributed to Edwards and Parry [4], Edwards [5], and other researchers who integrated RSM into their methodologies. They employed quadratic regression equations depicted on a three-dimensional surface to investigate congruence phenomena within organizational research, encompassing concepts such as fit, match, similarity, or agreement. Edwards and Parry [4], for instance, contributed formulae for testing and interpreting quadratic regression equations in the study of congruence within organizational research. These equations enable researchers to assess conceptual models that difference scores aim to represent. Due to the substantial contributions and expansions of this technique by these scholars, RSM has become widely employed in studying intricate relationships between pairs of predictor variables and an outcome variable in organizational studies [6–9] and beyond.

Barranti et al. [10] argue for several conceptual advantages offered by RSM. Firstly, it examines (mis)matches by modeling the association of input factors with a response using three-dimensional space. Secondly, RSM can address more nuanced questions than traditional approaches. Thirdly, it can be utilized to test whether one type of mismatch (e.g., an overestimate) is worse than another (e.g., an underestimate). Recognizing these strengths [4, 10, 11], RSM has been recommended for examining two-way interactions in moderated regression [12, 13]. This chapter aims to present a diverse array of alternative formulae for the slopes and curvatures of two curves in quadratic polynomial regression equations using response surface analysis, further advancing the applicability and understanding of this powerful methodology.

2. Existing formulas

The fundamental quadratic regression equation utilized in response surface methodology (RSM) is expressed as follows:

$$Z = b_0 + b_1X + b_2Y + b_3X^2 + b_4XY + b_5Y^2 + e, \tag{1}$$

where Z denotes the outcome variable (response), while X and Y are two predictors. Also, in this equation, b_0 is a constant, while b_1, b_2, b_3, b_4, and b_5 are linear, quadratic, and interaction terms. The outcome variable Z is regressed on two predictors X and Y, their respective square terms X^2 and Y^2, and their interaction XY.

Figure 1 visually represents a hypothetical response surface Z (depicted in blue) based on the quadratic regression equation mentioned above. The curve M (in red) is the intersected curve of the surface Z and plane ACC′A′ (Y = X). Plane BDD′B′ (Y = −X) intersects the surface Z at the curve M′ (dotted yellow curve). Line L (in black) is the tangent line to the intersected curve M at point P (0, 0, b_0) (in yellow). The origin point is O (0,0,0) and is marked in red.

According to Edwards and Parry [4] and Edwards (2002), the slopes and curvatures of the intersected curves M and M′ of the surface Z and two planes are important in exploring the nuanced relationships between two predictors and the outcome. One plane is the plane ACC′A′ (Y = X) (see **Figure 1**), while another is the plane BDD′B′

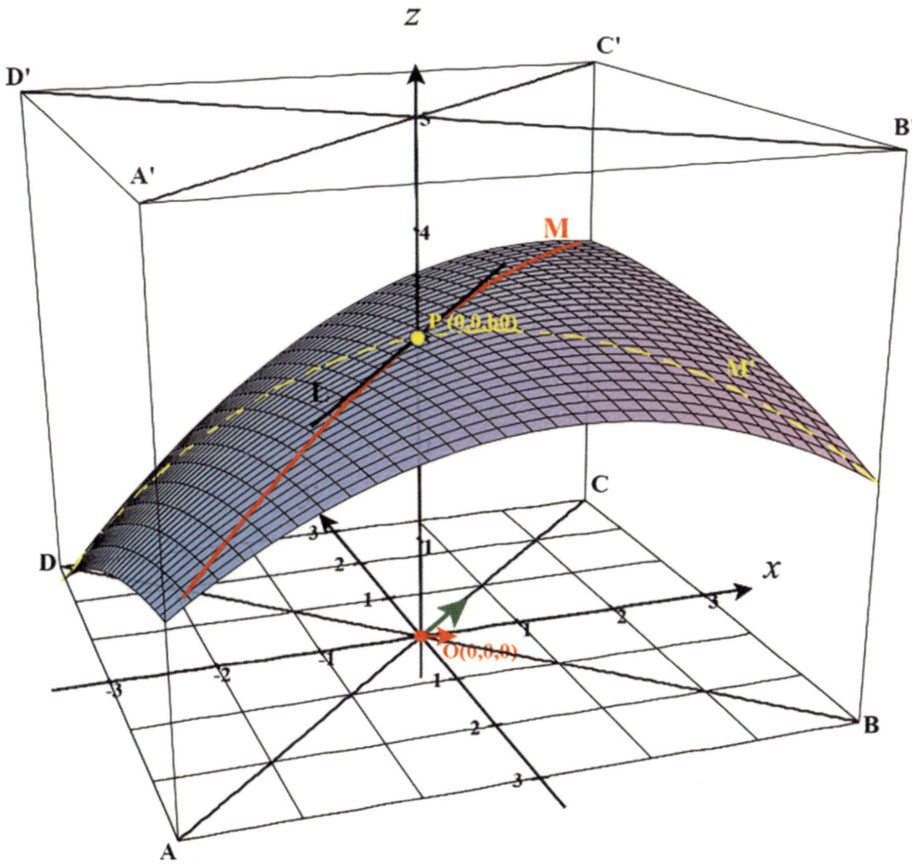

Figure 1.
A hypothetical response surface. The two predictor variables X and Y are located on the two axes at the bottom plane ABCD. The vertical axis is the Z-axis. Plane ACC'A' (Y = X) intersects the surface Z at the curve M. Plane BDD'B' (Y = −X) intersects the surface Z at the curve M'. The origin point is O (0,0,0).

(Y = −X), which is perpendicular to the plane ACC′A′ at Z-axis. To estimate the slope and curvature of intersection curve M, X is substituted for Y in the quadratic regression Eq. (1), which yields Eq. (2):

$$Z = b_0 + (b_1 + b_2)\, X + (b_3 + b_4 + b_5)\, X^2 + e, \tag{2}$$

Likewise, they derive quadratic regression Eq. (3) to estimate the slope and curvature of the curve M′,

$$Z = b_0 + (b_1 – b_2)\, X + (b_3 – b_4 + b_5)\, X^2 + e, \tag{3}$$

Then, they have

$$a_1 = b_1 + b_2 \tag{4}$$

$$a_2 = b_3 + b_4 + b_5 \tag{5}$$

$$a_3 = b_1 – b_2 \tag{6}$$

$$a_4 = b_3 – b_4 + b_5 \tag{7}$$

Where the term (b_3 + b_4 + b_5) is the curvature of the curve M and (b_1 + b_2) represents its slope at the point P (0, 0, b_0). Likewise, the term (b_3 – b_4 + b_5) represents the curvature of the curve M′, and (b_1 – b_2) is its slope at the point P (0, 0, b_0).

The formulas for calculating the significance tests of the curve values (a_1 through a_4) are also given as follows [13]:

$$\text{For variable } a_1 \text{ t} = \frac{(b1 + b2)}{\sqrt{(SEb1^2 + SEb2^2) + 2COV\ b1b2}} \tag{8}$$

$$\text{For variable } a_2 \text{ t} = \frac{(b3 + b4 + b5)}{\sqrt{(SEb3^2 + SEb4^2 + SEb5^2\ ) + 2COVb3b4 + 2COVb3b5 + 2COVb4b5}} \tag{9}$$

$$\text{For variable } a_3 \text{ t} = \frac{(b1 - b2)}{\sqrt{(SEb1^2 + SEb2^2) - 2COV\ b1b2}} \tag{10}$$

$$\text{For variable } a_4 \text{ t} = \frac{(b3 - b4 + b5)}{\sqrt{(SEb3^2 + SEb4^2 + SEb5^2\ ) - 2COVb3b4 + 2COVb3b5 - 2COVb4b5}} \tag{11}$$

Where SE = standard error, COV = covariance.

3. Alternative formulas

Presently, the formulas derived from Eqs. (2) and (3) play a pivotal role in various studies that leverage the response surface methodology (RSM) across a diverse array of fields. However, it is crucial to highlight a limitation in their applicability when transitioning to the context of three-dimensional geometry. Specifically, when substituting X (or −X) for Y in these equations, which are designed to represent curves M and M′ formed by the intersection of surface Z with planes ACC′A′ (Y = X) and BDD′B′ (Y = −X) a discrepancy arises.

Contrary to expectations, Eqs. (2) and (3) do not precisely depict curves M and M′ but rather represent two-dimensional projections onto the XOZ plane by the intersection curves M and M′. This distinction is particularly relevant when interpreting the slopes and curvatures of these projected curves in relation to the intersected curves within the three-dimensional space. Caution is warranted in utilizing these formulae as it may not always be appropriate or accurate to extrapolate the slopes and curvatures of projections to represent those of intersected curves, especially under specific circumstances.

This cautionary note emphasizes the need for a nuanced and context-specific approach when applying these formulas in three-dimensional geometric scenarios. Researchers and practitioners should be aware of the limitations associated with the projection of curves onto two-dimensional planes and exercise discretion in interpreting and applying the derived slopes and curvatures, ensuring that the chosen methodology aligns with the geometric intricacies of the study at hand. By acknowledging and addressing this caveat, the broader scientific community can enhance the precision and reliability of results when utilizing these formulas within the context of three-dimensional geometry.

The highlighted distinction holds paramount importance for researchers engaged in the intricate exploration of lines and curves within the realm of three-dimensional geometry. To achieve an accurate determination of slopes and curvatures concerning the intersected curves between surface Z and the two planes (Y = X and Y = −X), a more optimal approach is recommended. This involves leveraging the principles of directional derivatives and the rotation of axes, as commonly employed in multivariable calculus.

The application of directional derivatives facilitates a more nuanced examination of how the surface Z intersects with the planes, allowing researchers to discern the changes in the surface's behavior along specific directions. Additionally, the rotation of axes provides a transformative framework that aligns with the geometric intricacies of the study. By appropriately rotating the coordinate system, researchers can simplify the analysis and enhance the precision of calculations related to slopes and curvatures.

This advanced method not only ensures a more robust analysis but also aligns with the inherent complexities of three-dimensional geometry. It enables researchers to overcome the limitations associated with the two-dimensional projections mentioned earlier, providing a more accurate representation of the relationships between the curves and the intersecting planes. Embracing such sophisticated mathematical tools from multivariable calculus contributes to a higher level of precision and reliability in the study of intersected curves within three-dimensional spaces, ultimately enhancing the quality of research outcomes in this domain.

$$\text{For } Z = b_0 + b_1X + b_2Y + b_3X^2 + b_4XY + b_5Y^2 + e, \tag{12}$$

The directional derivative $D_L f(X_0, Y_0)$ equals the slope of the tangent line (L) to the intersected curve M of surface Z and the plane $ACC'A'$ (Y = X) at the point (X_0, $Y_0, f(X_0, Y_0)$). The formula to calculate the directional derivative of a curve at this point is:

$$D_L = \frac{\partial Z}{\partial l} = \frac{\partial Z}{\partial x}\ cos\ \theta + \frac{\partial Z}{\partial y}\ sin\ \theta \tag{13}$$

Since angle θ = 45°,

$$\begin{aligned} D_L = \frac{\partial Z}{\partial l} &= \frac{\partial Z}{\partial x}\ cos\ 45^{\circ} + \frac{\partial Z}{\partial y}\ sin\ 45^{\circ} \\ &= (b_1 + 2b_3X + b_4Y)\ cos\ 45^{\circ} + (b_2 + b_4X + 2b_5Y)\ sin\ 45^{\circ} \\ &= \frac{\sqrt{2}}{2}\left[(b_1 + b_2) + (2b_3 + b_4)\,X + (b_4 + 2b_5)\,Y\right] \end{aligned} \tag{14}$$

When X = 0 and Y = 0, the directional derivative $D_L f(0,0)$ equals $\frac{\sqrt{2}}{2}$ (b_1 + b_2) at the point P (0, 0, b_0). Obviously, the slope of the intersected curve M of surface Z and plane $ACC'A'$ (Y = X) is not (b_1 + b_2) at the point P (0, 0, b_0) as calculated by Edwards and Parry [4] and Edwards [5]. Similarly, the slope of the intersected curve M′ is $\frac{\sqrt{2}}{2}$ (b_1 – b_2) at the point P (0, 0, b_0), not (b_1 – b_2).

To capture both slope and curvature of the intersected curves, the rotation of axis is used to derive new rotated equations of intersected curves M and M′. The formula is provided as follows:

$$\begin{cases} Z = Z' \\ X = X' \cos\theta - Y' \sin\theta \\ Y = X' \sin\theta + Y' \cos\theta \end{cases} \tag{15}$$

For the intersected curve M, it is obtained by rotating 45° counterclockwise from the plane XOZ (X = 0); therefore, the new equation of curve M is

$$\begin{cases} Z = Z' \\ X = X' \cos 45^\circ - Y' \sin 45^\circ \\ Y = X' \sin 45^\circ + Y' \cos 45^\circ \end{cases} \tag{16}$$

Since $\sin 45^\circ = \cos 45^\circ = \frac{\sqrt{2}}{2}$, we have

$$\begin{cases} Z = Z' \\ X = \frac{\sqrt{2}}{2}(X' - Y') \\ Y = \frac{\sqrt{2}}{2}(X' + Y') \end{cases} \tag{17}$$

Thus,

$$\begin{aligned} Z' = {} & b_0 + b_1 * \frac{\sqrt{2}}{2}(X' - Y') + b_2 * \frac{\sqrt{2}}{2}(X' - Y') + b_3\left[\frac{\sqrt{2}}{2}(X' - Y')\right]^2 \\ & + b_4\left[\frac{\sqrt{2}}{2} * \frac{\sqrt{2}}{2}(X' - Y')(X' + Y')\right] + b_5\left[\frac{\sqrt{2}}{2}(X' + Y')\right]^2 \end{aligned} \tag{18}$$

Since in the intersected curve M, $Y' = 0$

$$Z' = b_0 + \frac{\sqrt{2}}{2}(b_1 + b_2) X' + \frac{1}{2}(b_3 + b_4 + b_5) X'^2 + e \tag{19}$$

Similarly, the new equation of intersected curve M′ is

$$Z' = b_0 + \frac{\sqrt{2}}{2}(b_1 - b_2) X' + \frac{1}{2}(b_3 - b_4 + b_5) X'^2 + e \tag{20}$$

Therefore,

$$a_1 = \frac{\sqrt{2}}{2}(b_1 + b_2) \tag{21}$$

$$a_2 = \frac{1}{2}(b_3 + b_4 + b_5) \tag{22}$$

$$a_3 = \frac{\sqrt{2}}{2}(b_1 - b_2) \tag{23}$$

$$a_4 = \frac{1}{2}(b_3 - b_4 + b_5) \tag{24}$$

The significance test t is calculated by using the following formula: $t = \frac{X-\mu}{SE}$.
In the case of the significance test of coefficient of X′, a_1

$$\begin{aligned} t &= \frac{a1}{SEa1} = \frac{\sqrt{2}}{2}(b_1 + b_2)/\frac{\sqrt{2}}{2}\sqrt{(SEb1^2 + SEb2^2) + 2COV\ b1b2} \\ &= (b_1 + b_2)/\sqrt{(SEb1^2 + SEb2^2) + 2COV\ b1b2} \end{aligned} \tag{25}$$

Similarly,

$$\text{for variable } a_2 \text{ t} = \frac{(b3 + b4 + b5)}{\sqrt{(SEb3^2 + SEb4^2 + SEb5^2) + 2COVb3b4 + 2COVb3b5 + 2COVb4b5}} \tag{26}$$

$$\text{for variable a3 t} = \frac{(b1 - b2)}{\sqrt{(SEb1^2 + SEb2^2) - 2COV\ b1b2}} \tag{27}$$

$$\text{for variable } a_4 \text{ t} = \frac{(b3 - b4 + b5)}{\sqrt{(SEb3^2 + SEb4^2 + SEb5^2) - 2COVb3b4 + 2COVb3b5 - 2COVb4b5}} \tag{28}$$

Upon close examination of the aforementioned analysis, it becomes apparent that the values of coefficients (a1, a2, a3, and a4) and their corresponding standard errors exhibit notable discrepancies compared to the values reported by Edwards and Parry [4] and Edwards [5]. Interestingly, the t values, however, remain consistent, retaining their original significance due to a canceling-out effect.

The observed differences in coefficient values and standard errors suggest a departure from the previously established models proposed by Edwards and Parry [4] and Edwards [5]. It is essential to scrutinize these distinctions meticulously to discern the potential implications for the interpretation and generalization of the quadratic polynomial regression equations under consideration. Despite the dissimilarities in coefficient values and standard errors, the unchanged t values indicate that the significance levels associated with the coefficients remain unaltered. This underscores the robustness of the statistical significance of the individual coefficients, emphasizing their reliability even in the face of differing parameter estimates.

4. An empirical example

In order to elucidate the differentiation between the two previously mentioned formulae, we conducted an empirical analysis using the initial dataset derived from polynomial regression studies carried out by Qiu, Dooley, and Xie [12]. The referenced paper encompasses a comprehensive exploration involving two distinct studies, each contributing valuable insights to the field.

The dataset for the first study was meticulously gathered from front-line employees within a prominent restaurant chain. The primary aim of this study was to delve into the intricate dynamics of the interaction between servant leadership and self-efficacy, examining its profound impact on service quality within the context of the hospitality industry. This research undertaking reflects a targeted effort to understand the nuanced relationships and interplay between leadership styles and individual efficacy in a service-oriented setting, specifically within the dynamic environment of a restaurant chain.

By selecting this particular dataset, we sought to ground our exploration in a real-world scenario, one that encapsulates the complexities and intricacies of human interactions within the hospitality sector. The utilization of data from front-line employees adds a practical dimension to our analysis, enabling us to draw meaningful conclusions about the potential variations in slopes and curvatures of the two curves within quadratic polynomial regression equations. Through this empirical approach, we aim to not only highlight the distinctions between the formulae but also to provide valuable insights that contribute to the broader understanding of how these mathematical models can be applied and interpreted in the realm of organizational research, particularly within the hospitality industry.

The polynomial regression equation they derived from the response surface analysis was expressed as $Z = 3.481 + 0.551X + 0.227Y - 0.234XY + 0.119Y^2 + e$. Based on Eq. (1) described above, it can be easily seen that $b_0 = 3.481$, $b_1 = 0.551$, $b_2 = 0.227$, $b_3 = 0$, $b_4 = -0.234$, $b_5 = 0.119$. They then calculated the values of a_1, a_2, a_3, and a_4 as follows: $a_1 = 0.778^{**}$, $a_2 = 0.051$, $a_3 = 0.324^{**}$, and $a_4 = 0.417^{**}$ ($^{**}p < 0.01$) following the formulae provided by Edwards and Parry [4]. However, if readers use the correct formulas presented previously, it can be derived that $a_1 = 0.550^{**}$, $a_2 = 0.026$, $a_3 = 0.229^{**}$, and $a_4 = 0.209^{**}$ ($^{**}p < 0.01$). Here, *t* scores were not calculated because the significance levels of the coefficients remain unchanged.

For a more detailed examination of the statistical results, including t scores and additional relevant information, we recommend referring to Table 2 in the paper authored by Qiu, Dooley, and Xie [12]. This table likely provides a comprehensive summary of the regression analysis, offering insights into the specific values of coefficients, standard errors, t scores, and corresponding significance levels. By consulting this table, readers can gain a more nuanced understanding of the statistical nuances and intricacies involved in the analysis, thereby enhancing the overall comprehension of the research findings presented in the paper.

5. Conclusions

This paper utilized directional derivatives and rotations of axes, concepts from multivariable calculus, to derive an alternative set of formulae for calculating the slopes and curvatures of two curves within quadratic polynomial regression equations through response surface analysis. To emphasize, the accurate formulas for a_1, a_2, a_3, and a_4 are as follows:

$$a_1 = \frac{\sqrt{2}}{2}(b_1 + b_2) \tag{29}$$

$$a_2 = \frac{1}{2}(b_3 + b_4 + b_5) \tag{30}$$

$$a_3 = \frac{\sqrt{2}}{2}(b_1 - b_2) \tag{31}$$

$$a_4 = \frac{1}{2}\left(b_3 – b_4 + b_5\right). \tag{32}$$

The formulas for the significance of a_1, a_2, a_3, and a_4 are as follows:

$$\text{For variable } a_1 \; t_{a1} = \frac{(b1 + b2)}{\sqrt{(SEb1^2 + SEb2^2) + 2COV\ b1b2}} \tag{33}$$

$$\text{For variable } a_2 \; t_{a2} = \frac{(b3 + b4 + b5)}{\sqrt{(SEb3^2 + SEb4^2 + SEb5^2\ ) + 2COVb3b4 + 2COVb3b5 + 2COVb4b5}} \tag{34}$$

$$\text{For variable } a_3 \; t_{a3} = \frac{(b1 - b2)}{\sqrt{(SEb1^2 + SEb2^2) - 2COV\ b1b2}} \tag{35}$$

$$\text{For variable } a_4 \; t_{a4} = \frac{(b3 - b4 + b5)}{\sqrt{(SEb3^2 + SEb4^2 + SEb5^2\ ) - 2COVb3b4 + 2COVb3b5 - 2COVb4b5}} \tag{36}$$

Our primary goal is to systematically present a comprehensive assortment of mathematically rigorous formulae, poised to significantly enhance our comprehension of the slopes and curvatures inherent in quadratic polynomial regression equations through the prism of response surface analysis. The pursuit of this objective becomes imperative in light of the intricate mathematical intricacies entwined with such investigative endeavors. Our intent is not only to unveil theoretical frameworks but also to ground them in practicality, as evidenced by the empirical example we present. Through this illustration, we seek to shed light on the nuanced distinctions in slopes between the two previously outlined formulae, offering researchers a tangible and applicable perspective.

In elaborating on our empirical example, we emphasize its role as a pragmatic demonstration, providing a clear and tangible portrayal of how the theoretical constructs can be translated into real-world scenarios. By elucidating the disparities in slopes between the two curves within quadratic polynomial regression equations, we aim to demystify complex mathematical concepts and encourage a deeper understanding of the underlying principles.

Furthermore, we extend an invitation to the broader research community, urging scholars across diverse fields to adopt and integrate this newly proposed analytical tool into their investigations. The application of response surface analysis to explore the slopes and curvatures of two curves within quadratic polynomial regression equations holds promising potential for unlocking insights across various disciplines. By advocating for the widespread use of this innovative approach, we aspire to catalyze advancements in knowledge and contribute to the refinement of methodologies employed in quantitative analyses within respective research domains.

Author details

Shaoping Qiu[1]* and Kaizhe (Kaiser) Qiu[2]

1 Department of Leadership Studies, Louisiana State University, Shreveport, USA

2 College of Science and Engineering, University of Minnesota, Twin Cities, USA

*Address all correspondence to: qap5415@gmail.com

References

[1] Myers RH, Montgomery DC, Anderson-Cook CM. Response Surface Methodology: Process and Product Optimization Using Designed Experiments. New York, NY: John Wiley & Sons; 2016

[2] Box GEP, Wilson KB. On the experimental attainment of optimum conditions. Journal of the Royal Statistical Society, Series B. 1951;**13**(1): 1-45. DOI: 10.1111/j.2517-6161.1951. tb00067.x

[3] Khuri AI, Cornell JA. Response Surfaces: Designs and Analyses. New York, NY: Routledge; 2018

[4] Edwards JR, Parry ME. On the use of polynomial regression equations as an alternative to difference scores in organizational research. Academy of Management Journal. 1993;**36**(6): 1577-1613

[5] Edwards JR. The study of congruence in organizational behavior research: Critique and a proposed alternative. Organizational Behavior and Human Decision Processes. 1994;**58**(1):51-100

[6] Bernerth JB, Carter MZ, Cole MS. The (in) congruence effect of leaders' narcissism identity and reputation on performance: A socioanalytic multistakeholder perspective. Journal of Applied Psychology. 2021;**107**(10):1725-1742. DOI: 10.1037/apl0000974

[7] Briker R, Walter F, Cole MS. The consequences of (not) seeing eye-to-eye about the past: The role of supervisor-team fit in past temporal focus for supervisors' leadership behavior. Journal of Organizational Behavior. 2020;**41**(3): 244-262. DOI: 10.1002/job.2416

[8] Graham KA, Mawritz MB, Dust SB, Greenbaum RL, Ziegert JC. Too many cooks in the kitchen: The effects of dominance incompatibility on relationship conflict and subsequent abusive supervision. The Leadership Quarterly. 2019;**30**(3):351-364. DOI: 10.1016/j.leaqua.2018.12.003

[9] Grant AM. Rocking the boat but keeping it steady: The role of emotion regulation in employee voice. Academy of Management Journal. 2013;**56**(6): 1703-1723. DOI: 10.5465/amj.2011.0035

[10] Barranti M, Carlson EN, Côté S. How to test questions about similarity in personality and social psychology research: Description and empirical demonstration of response surface analysis. Social Psychological and Personality Science. 2017;**8**(4):465-475

[11] Humberg S, Nestler S, Back MD. Response surface analysis in personality and social psychology: Checklist and clarifications for the case of congruence hypotheses. Social Psychological and Personality Science. 2019;**10**(3):409-419. DOI: 10.1177/1948550618757600

[12] Qiu S, Dooley LM, Xie L. How servant leadership and self-efficacy interact to affect service quality in the hospitality industry: A polynomial regression with response surface analysis. Tourism Management. 2020;**78**:104051 DOI: 10.1016/j.tourman.2019.104051

[13] Shanock LR, Baran BE, Gentry WA, Pattison SC, Heggestad ED. Polynomial regression with response surface analysis: A powerful approach for examining moderation and overcoming limitations of difference scores. Journal of Business and Psychology. 2010;**25**(4): 543-554. DOI: 10.1007/s10869-010-9183-4

Chapter 5

Constant Block-Size with Constant Sum-Block Partially Balanced Incomplete Block Designs

Babatunde L. Adeleke, Gabriel O. Adebayo and Kazeem A. Osuolale

Abstract

This chapter delves into the introduction, methodology, and application of Constant Block-Size with Constant Sum-Block Partially Balanced Incomplete Block Designs (CBS-CSB PBIBDs) as an innovative class of experimental designs. The study defines key concepts, including constant block-size, constant block-sum, and constant sum-block, and outlines conditions for implementing CBS-CSB PBIBDs. The research presents experimental designs for various scenarios, such as t = 21, k = 3; t = 15, k = 3; t = 27, k = 3, and illustrates their respective cases through line charts. Notably, the study demonstrates that as the number of replicates (r) increases, the efficiency factors also increase, and when r = 1, efficiency (E) equals 0. In conclusion, CBS-CSB PBIBDs emerge as a valuable tool for experimental prioritization, offering a new perspective on partially balanced incomplete block designs. The research highlights the significance of CBS-CSB PBIBDs across diverse fields of experimentation. The findings equip readers with the ability to define, design, and analyze experiments using CBS-CSB PBIBDs, while acknowledging challenges and limitations associated with this approach. This study contributes to the broader understanding of experimental design and provides a foundation for future research directions in this innovative field.

Keywords: experimental design, partially balanced incomplete block designs, constant block-size, constant sum-block, prioritization of treatments

1. Introduction

Constant Block-Size with Constant Sum-Block Partially Balanced Incomplete Block Designs, CBS-CSB PBIBDs is a new class of partially balanced incomplete block designs that is use when it requires prioritizing the treatments combination. This new class of the design is significant in every aspect of experiment in physical, life, clinical, engineering, education, social science, psychology, pharmaceutical etc. once there is need to prioritize the treatment combination then CBS-CSB PBIBDs will be used.

The goal of this chapter is to define constant block-size partially balanced incomplete block designs and constant sum-block partially balanced incomplete block designs, design experiments using constant block-size with constant sum-block partially balanced incomplete block designs, CBS-CSB PBIBDs, state the Conditions for Constant Block-Size with Constant Sum-Block Partially Balanced Incomplete Block Designs, CBS-CSB PBIBDs, perform simple analysis on CBS-CSB PBIBDs and state some challenges and limitation on CBS-CSB PBIBDs.

In the remaining part of this chapter, we will discuss the background and theory, design principles and methodology, data analysis, discussion of results and conclusion of study's findings.

1.1 Background and theory

The optimum design is balanced incomplete block designs (BIBDs) which gives 100% efficiency factor, e = 1. But the shortcoming of BIBDs leads to the introduction of partially balanced incomplete block designs (PBIBDs) introduced by the authors in [1]. This chapter introduces partially balanced incomplete block designs and its Efficiency Factors but its limitation was that it took into consideration the intra-block information only. The important properties of partially balanced incomplete block designs with two associate classes with both the intra-block and inter-block analysis in a comprehensive was introduced by the author [2]. Analysis of partially balanced incomplete block designs using simple square and rectangular lattices [3], construction of partially balanced incomplete block designs with two treatments per block and less or equal to ten (10) replications of each treatment [4], the class of partially balanced incomplete block designs with more than two associate classes comprehensively [5], m-classes partially balanced incomplete block designs and association scheme [6] and the table of 2-associate classes of partially balanced incomplete block designs were constructed [7].

There are several patterns and strategies in the construction of PBIBDs for the purpose of estimating higher efficiency factors. Some of the strategies are complicated and complex in designs and the efficiency factors is not satisfactory and comprehensive. The PBIBDs with higher associate classes were constructed by authors in [8–10]. Triangular and four associate classes PBIBDs with two replicates using dualization method were constructed by other scholars that contributed to the construction of PBIBDs with their different and respective pattern and strategies [11–14]. In the physical and life experiment, there may be need to prioritize some treatment combinations. It was observed that development of PBIBDs through the incorporation of a constant block-sum strategy [15, 16] and later, in a subsequent work [17] delved deeper into and expanded upon this strategy in the construction of PBIBDs. This design changed the new pattern and strategy towards the construction of PBIBDs by authors cited in [18–22]. The limitations of the design was that $r = 1$, $\lambda_1 = 1$, $\lambda_2 = 0$. If $r = 1$, then $E = 0$ where 'E' is the efficiency factor, 'r' is the number of replicate and 'λ' is the number of treatment pair; this does not fit for analysis. To design an experiment, $r \geq 2$, for r is positive integer. Instead, the design may be categorized into two or more to fit for analysis [23]. The key concepts for this work are Partially Balanced Incomplete Block Design; Constant Block-Size; Constant Sum-Block; Constant Block-Sum; Efficiency Factors.

2. Design principles and methodology

In this section, the paper defines, give conditions and design CBS-CSB PBIBDs.

2.1 Definition of terms

This sub-section defines Constant Block-Size, Constant Block-Sum and Constant Sum-Block of partially balanced incomplete block designs;

Definition 1: A design is said to have a constant block-size if the number of experimental unit, k, are the same irrespective of the number of blocks, b. For example;

- t = 21, k = 3, r = 1, b = 7
- t = 21, k = 3, r = 2, b = 14
- t = 21, k = 3, r = 3, b = 21

Definition 2: A design is said to be constant block-sum if the number of blocks, b, are the same, b $= \frac{t}{k}$, and the sum of each corresponding number of experimental units are the same, irrespective of the number of treatments and replicates [24]. It is determined by the sum of all the treatments divided by number of blocks, CBS $= \frac{\sum t}{b}$. Furthermore, Khattree model the design as:

t = pq, k = min (p,q), b = max (p,q), r = 1, $\lambda_1 = 1$, $\lambda_2 = 0$ where p and q must be odd numbers.

For example;

$$\mathrm{t} = 21, \mathrm{k} = 3, \mathrm{r} = 1, \mathrm{b} = 7, \lambda_1 = 1, \lambda_2 = 0, \mathrm{n}_1 = 2, \mathrm{n}_2 = 18 \tag{1}$$

$$\mathrm{CSB} = \frac{1 + 2 + \ldots + 21}{7} = \frac{231}{7} = 33. \tag{2}$$

Table 1 shows a Constant Block-Sum table, where each row (b1, b2, b3, etc.) represents a block, and the numbers within each block represent the treatments. The key feature of this table is that the sum of values in each block is constant and equal to 33.

Definition 3: A design is said to be a constant sum-block, if the following conditions exist;

- There is constant block size.
- Each corresponding blocks has constant sum.
- The number of blocks are not the same, b $= \frac{tr}{k}$

				Sum
b1	1	11	21	33
b2	2	15	16	33
b3	3	10	20	33
b4	4	12	17	33
b5	5	9	19	33
b6	6	13	14	33
b7	7	8	18	33

Table 1.
Constant block-sum table.

- The number of treatments, t, are constant.
- The number of replicates, r, are at least one, $r \geq (1)$

For example;

$$t = 21, k = 3, r = 2, b = 14, \lambda_1 = 1, \lambda_2 = 0, n_1 = 4, n_2 = 16.$$

In **Table** 2, each row represents a block labeled as b1, b2, b3, and so on. The numbers within each block represent different treatments. For example, in block b1, the treatments are 1, 11, 21, and the sum of these treatments is 33. The same pattern follows for the other blocks. The sum of treatments in each block is the same and equal to 33. This is evident by the "Sum" column at the end of each row, indicating that the values in each row (treatments within a block) add up to 33. This design, with a constant sum within each block is used in experimental design to ensure balance and control for variability. It allows for a systematic distribution of treatments across different blocks while maintaining a consistent sum within each block.

2.2 Conditions for constant block-size with constant sum-block partially balanced incomplete block designs, CBS-CSB PBIBDs

There are conditions for CBS-CSB PBIBDs;

- The treatments combinations must be prioritized, that is, there must be number of zero associates, $\lambda_2 = 0$.

				Sum
b1	1	11	21	33
b2	2	15	16	33
b3	3	10	20	33
b4	4	12	17	33
b5	5	9	19	33
b6	6	13	14	33
b7	7	8	18	33
b8	1	12	20	33
b9	2	14	17	33
b10	3	9	21	33
b11	4	11	18	33
b12	5	13	15	33
b13	6	8	19	33
b14	7	10	16	33

Table 2.
Constant sum-block table.

- The maximum number of replicates must be necessarily determine from treatment one.
- The number of treatments and block sizes remain fixed.

For example;
The designs;

$$t = 21, k = 3, \lambda_2 = 0,$$

The three numerical values in each row in **Table 3** represent different treatments or conditions. For example, in the first row (SN = 1), the values are 1, 21, and 11 and their sum is 33. Similarly, the sum of values in each subsequent row is consistently 33. The key feature here, as in the previous tables, is that the sum of the numerical values in each row is constant and equal to 33. This ensures a balanced distribution of the treatments or conditions. In this context, the table structure represents a specific design where the sum of values in each row is held constant. The associate for treatment 1 is given in **Table 4**.

Table 4 associates different sets of values with two conditions, denoted as λ_1 and λ_2, for a specific treatment labeled as T. The row labeled as T corresponds to a specific treatment condition while λ1 and λ2 are conditions or parameters associated with the treatment. For treatment 1, the associated values are presented under λ1 = 1 and λ2 = 0. Under each condition (λ1 and λ2), there are sets of values associated with the treatment. For example, under λ1 = 1, the associated values are 11, 21, 12, 20, 13, 19, 14, 18, 15, 17. Under λ2 = 0, the associated values are 2, 3, 4, 5, 6, 7, 8, 9, 10, 16. **Table 4** shows that $1 < r \leq 5$ and $\lambda_2 = 0$. This met the conditions for CBS-CSB PBIBDs.

2.3 Case studies on the design constant block-size with constant sum-block partially balanced incomplete block designs, CBS-CSB PBIBDs

This section shows the construction of CBS-CSB PBIBDs. The researcher and the user will determine the number of replicates based on the priority of the treatment combination and the cost of the experiment (**Table 5**). This means $1 < r \leq \max(r)$.

SN				Sum
1	1	21	11	33
2	1	20	12	33
3	1	19	13	33
4	1	18	14	33
5	1	17	15	33

Table 3.
Maximum number of replicates table.

T	$\lambda_1 = 1$	$\lambda_2 = 0$
1	11, 21,12,20,13,19,14,18,15,17	2,3,4,5,6,7,8,9,10,16

Table 4.
Associate table for treatment 1.

				Sum
b1	1	11	21	33
b2	2	15	16	33
b3	3	10	20	33
b4	4	12	17	33
b5	5	9	19	33
b6	6	13	14	33
b7	7	8	18	33

Table 5.
Case 1 design.

2.3.1 Construction of design 1

The design t = 21, k = 3, λ_2 = 0 will be constructed in five cases since r = 5.

Case 1: t = 21, k = 3, r = 1, b = 7, λ_1 = 1, λ_2 = 0, n_1 = 2, n_2 = 18.

The table represents the design with 7 blocks (b1 to b7), and each block contains 3 treatments (k = 3). The values in each block are arranged in a way that the sum of treatments within each block is constant and equal to 33. This design follows a structure for the Constant Block-Sum table provided earlier, with the same sum of treatments in each block.

$$P_1 = \begin{bmatrix} 1 & 0 \\ 0 & 18 \end{bmatrix}, P_2 = \begin{bmatrix} 0 & 2 \\ 2 & 15. \end{bmatrix}$$

$$E_1 = 0, E_2 = 0, E = 0$$

The efficiency factor (E) is calculated using R software based on a method proposed by authors in [25]. **Table 6** associates each treatment (T) with sets of values for the 1st and 2nd conditions. For example, treatment 1 (T = 1) is associated with values 11 and 21 under the 1st condition, and with values 2, 3, 4, 5, 6, 7, 8, 9, 10, 12, 13, 14, 15, 16, 17, 18, 19, 20 under the 2nd condition. The structure of the table indicates how each treatment is linked to specific sets of values under different conditions. This indicates that the efficiency is not directly given but needs to be calculated using specific software and methodology. Case 2 designs are shown in **Tables** 7 and **8**, respectively.

Case 2: t = 21, k = 3, r = 2, b = 14, λ_1 = 1, λ_2 = 0, n_1 = 4, n_2 = 16

$$P_1 = \begin{bmatrix} 1 & 2 \\ 2 & 14 \end{bmatrix}, P_2 = \begin{bmatrix} 0 & 4 \\ 4 & 11 \end{bmatrix}$$

$$E_1 = 0.6667, E_2 = 0.4444, E = 0.4762$$

Each row in **Table 7** represents a block, and there are 14 blocks in total. The numbers within each block represent different treatments. The sum of treatments in each block is constant and equal to 33, maintaining a similar structure to the design in Case 1. **Table 8** associates each treatment (T) with sets of values for the 1st and 2nd conditions. For instance, treatment 1 (T = 1) is associated with values 11, 21, 12, and 20

T	1st	2nd
1	11, 21	2,3,4,5,6,7,8,9,10,12,13,14,15,16,17,18,19,20
2	15,16	1,3,4,5,6,7,8,9,10,11,12,13,14,17,18,19,20,21
3	10, 20	1,2,4,5,6,7,8,9,11,12,13,14,15,16,17,18,19,21
4	12, 17	1,2,3,5,6,7,8,9,10,11,13,14,15,16,18,19,20,21
5	9, 19	1,2,3,4,6,7,8,10,11,12,13,14,15,16,17,18,20,21
6	13, 14	1,2,3,4,5,7,8,9,10,11,12,15,16,17,18,19,20,21
7	8, 18	1,2,3,4,5,6,9,10,11,12,13,14,15,16,17,19,20,21
8	7, 18	1,2,3,4,5,6,9,10,11,12,13,14,15,16,17,19,20,21
9	5, 19	1,2,3,4,6,7,8,10,11,12,13,14,15,16,17,18,20,21
10	3, 20	1,2,4,5,6,7,8,9,11,12,13,14,15,16,17,18,19,21
11	1, 21	2,3,4,5,6,7,8,9,10,12,13,14,15,16,17,18,19,20
12	4,17	1,2,3,5,6,7,8,9,10,11,13,14,15,16,18,19,20,21
13	6,14	1,2,3,4,5,7,8,9,10,11,12,15,16,17,18,19,20,21
14	6,13	1,2,3,4,5,7,8,9,10,11,12,15,16,17,18,19,20,21
15	2,16	1,3,4,5,6,7,8,9,10,11,12,13,14,17,18,19,20,21
16	2,15	1,3,4,5,6,7,8,9,10,11,12,13,14,17,18,19,20,21
17	4,12	1,2,3,5,6,7,8,9,10,11,13,14,15,16,18,19,20,21
18	7,8	1,2,3,4,5,6,9,10,11,12,13,14,15,16,17,19,20,21
19	5,9	1,2,3,4,6,7,8,10,11,12,13,14,15,16,17,18,20,21
20	3,10	1,2,4,5,6,7,8,9,11,12,13,14,15,16,17,18,19,21
21	1,11	2,3,4,5,6,7,8,9,10,12,13,14,15,16,17,18,19,20

Table 6.
Case 1 associates.

under the 1st condition, and with values 2, 3, 4, 5, 6, 7, 8, 9, 10, 13, 14, 15, 16, 17, 18, 19 under the 2nd condition.

Case 3: $t = 21$, $k = 3$, $r = 3$, $b = 21$, $\lambda_1 = 1$, $\lambda_2 = 0$, $n_1 = 6$, $n_2 = 14$

$$P_1 = \begin{bmatrix} 2 & 3 \\ 3 & 11 \end{bmatrix}, P_2 = \begin{bmatrix} 2 & 4 \\ 4 & 9 \end{bmatrix}$$

$$E_1 = 0.7111, E_2 = 0.5926, E = 0.6238$$

The table represents the design with 21 blocks (b1 to b21), and each block contains 3 treatments (k = 3). The values in each block are arranged in a way that the sum of treatments within each block is constant and equal to 33. This design follows a structure similar to the Constant Block-Sum tables provided earlier, with the same sum of treatments in each block (**Table 9**). **Table 10** is similar to the previous associates tables. It associates each treatment (T) with sets of values for the 1st and 2nd conditions. For example, treatment 1 (T = 1) is associated with values 11, 21, 12, 20, 13, and 19 under the 1st condition, and with values 2, 3, 4, 5, 6, 7, 8, 9, 10, 14, 15, 16, 17, 18 under the 2nd condition.

				Sum
b1	1	11	21	33
b2	2	15	16	33
b3	3	10	20	33
b4	4	12	17	33
b5	5	9	19	33
b6	6	13	14	33
b7	7	8	18	33
b8	1	12	20	33
b9	2	14	17	33
b10	3	9	21	33
b11	4	11	18	33
b12	5	13	15	33
b13	6	8	19	33
b14	7	10	16	33

Table 7.
Case 2 designs.

Case 4: t = 21, k = 3, r = 4, b = 21, λ_1 = 1, λ_2 = 0, n_1 = 8, n_2 = 12

$$P_1 = \begin{bmatrix} 3 & 4 \\ 4 & 8 \end{bmatrix}, P_2 = \begin{bmatrix} 5 & 3 \\ 3 & 8 \end{bmatrix}$$

$$E_1 = 0.713, E_2 = 0.6417, E = 0.6684$$

Table 11 represents the design for Case 4, with specific values assigned to each block. Each row corresponds to a block (b1 to b28), and within each block, there are different treatments arranged in a way that the sum of treatments is constant and equal to 33. **Table 12** is similar to the previous associate tables. It associates each treatment (T) with sets of values for the 1st and 2nd conditions. For example, treatment 1 (T = 1) is associated with values 11, 21, 12, 20, 13, 19, 14, and 18 under the 1st condition, and with values 2, 3, 4, 5, 6, 7, 8, 9, 10, 15, 16, 17 under the 2nd condition. The notations P1 and P2 are partitions with specific characteristics. The **E1 = 0.713, E2 = 0.6417, E = 0.6684** values represent efficiency factors. In experimental design, efficiency is a measure of how well an experimental design utilizes the available resources. Here, E1 and E2 are specific efficiency values for P1 and P2, and E is an overall efficiency factor.

Case 5: t = 21, k = 3, r = 5, b = 21, λ_1 = 1, λ_2 = 0, n_1 = 10, n_2 = 10

$$P_1 = \begin{bmatrix} 4 & 5 \\ 5 & 5 \end{bmatrix}, P_2 = \begin{bmatrix} 8 & 2 \\ 2 & 7 \end{bmatrix}$$

$$E_1 = 0.7077, E_2 = 0.6571, E = 0.6815$$

Table 13 presents a Case 5 design. It represents the design for Case 5, with specific values assigned to each block. Each row corresponds to a block (b1 to b35), and within each block, there are different treatments arranged in a way that the sum of

T	1st	2nd
1	11, 21,12,20	2,3,4,5,6,7,8,9,10,13,14,15,16,17,18,19
2	15,16,14,17	1,3,4,5,6,7,8,9,10,11,12,13,18,19,20,21
3	10, 20,9,21	1,2,4,5,6,7,8,11,12,13,14,15,16,17,18,19
4	12, 17,11,18	1,2,3,5,6,7,8,9,10,13,14,15,16,19,20,21
5	9, 19,13,15	1,2,3,4,6,7,8,10,11,12,14,16,17,18,20,21
6	13, 14,8,19	1,2,3,4,5,7,9,10,11,12,15,16,17,18,20,21
7	8, 18,10,16	1,2,3,4,5,6,9,11,12,13,14,15,17,19,20,21
8	7, 18,6,19	1,2,3,4,5,9,10,11,12,13,14,15,16,17,20,21
9	5, 19,3,21	1,2,4,6,7,8,10,11,12,13,14,15,16,17,18,20
10	3, 20,7,16	1,2,4,5,6,8,9,11,12,13,14,15,17,18,19,21
11	1, 21,4,18	2,3,5,6,7,8,9,10,12,13,14,15,16,17,19,20
12	4,17,1,20	2,3,5,6,7,8,9,10,11,13,14,15,16,18,19,21
13	6,14,5,15	1,2,3,4,7,8,9,10,11,12,16,17,18,19,20,21
14	6,13,2,17	1,3,4,5,7,8,9,10,11,12,15,16,18,19,20,21
15	2,16,5,13	1,3,4,6,7,8,9,10,11,12,14,17,18,19,20,21
16	2,15,7,10	1,3,4,5,6,8,9,11,12,13,14,17,18,19,20,21
17	4,12,2,14	1,3,5,6,7,8,9,10,11,13,15,16,18,19,20,21
18	7,8,4,11	1,2,3,5,6,9,10,12,13,14,15,16,17,19,20,21
19	5,9,6,8	1,2,3,4,7,10,11,12,13,14,15,16,17,18,20,21
20	3,10,1,12	2,4,5,6,7,8,9,11,13,14,15,16,17,18,19,21
21	1,11,3,9	2,4,5,6,7,8,10,12,13,14,15,16,17,18,19,20

Table 8.
Case 2 associates.

treatments is constant and equal to 33. **Table 14** contained Case 5 associates. This table shows the associations between the treatments in Case 5. Each row represents a treatment (T1 to T21), and the "1st" and "2nd" columns indicate which other treatments are associated with the corresponding treatment. For example, the first row indicates that for Treatment 1 (T1), the associated treatments in the 1st group are 11, 21, 12, 20, 13, 19, 14, 18, 15, 17, and in the 2nd group are 2, 3, 4, 5, 6, 7, 8, 9, 10, 16. Similarly, each subsequent row provides information about the associations for the corresponding treatment. The efficiency values (E1, E2, and E) are also provided as 0.7077, 0.6571, and 0.6815, respectively (**Figure 1**).

2.3.2 Construction of design 2

The design $t = 15$, $k = 3$, $\lambda_2 = 0$ will be constructed in five cases since $r = 4$.

Case 1: $t = 15$, $k = 3$, $r = 1$, $b = 10$, $\lambda_1 = 1$, $\lambda_2 = 0$, $n_1 = 2$, $n_2 = 12$

$$P_1 = \begin{bmatrix} 1 & 0 \\ 0 & 12 \end{bmatrix}, P_2 = \begin{bmatrix} 0 & 2 \\ 2 & 9 \end{bmatrix}$$

				Sum
b1	1	11	21	33
b2	2	15	16	33
b3	3	10	20	33
b4	4	12	17	33
b5	5	9	19	33
b6	6	13	14	33
b7	7	8	18	33
b8	1	12	20	33
b9	2	14	17	33
b10	3	9	21	33
b11	4	11	18	33
b12	5	13	15	33
b13	6	8	19	33
b14	7	10	16	33
b15	1	13	19	33
b16	2	11	20	33
b17	3	14	16	33
b18	4	8	21	33
b19	5	10	18	33
b20	6	12	15	33
b21	7	9	17	33

Table 9.
Case 3 design.

$$E_1 = 0, E_2 = 0, E = 0$$

Case 2

$$t = 15, k = 3, r = 2, b = 10, \lambda_1 = 1, \lambda_2 = 0, n_1 = 4, n_2 = 10$$

$$P_1 = \begin{bmatrix} 1 & 2 \\ 2 & 8 \end{bmatrix}, P_2 = \begin{bmatrix} 2 & 2 \\ 2 & 7 \end{bmatrix}$$

$$E_1 = 0.75, E_2 = 0.6, E = 0.6364$$

Case 3

$$t = 15, k = 3, r = 3, b = 15, \lambda_1 = 1, \lambda_2 = 0, n_1 = 6, n_2 = 8$$

$$P_1 = \begin{bmatrix} 2 & 3 \\ 3 & 5 \end{bmatrix}, P_2 = \begin{bmatrix} 4 & 2 \\ 2 & 5 \end{bmatrix}$$

$$E_1 = 0.7302, E_2 = 0.6389, E = 0.6751$$

T	1st	2nd
1	11, 21,12,20,13,19	2,3,4,5,6,7,8,9,10,14,15,16,17,18
2	15,16,14,17,11,20	1,3,4,5,6,7,8,9,10,12,13,18,19,21
3	10, 20,9,21,14,16	1,2,4,5,6,7,8,11,12,13,15,17,18,19
4	12, 17,11,18,8,21	1,2,3,5,6,7,9,10,13,14,15,16,19,20
5	9, 19,13,15,10,18	1,2,3,4,6,7,8,11,12,14,16,17,20,21
6	13, 14,8,19,12,15	1,2,3,4,5,7,9,10,11,16,17,18,20,21
7	8, 18,10,16,9,17	1,2,3,4,5,6,11,12,13,14,15,19,20,21
8	7, 18,6,19,4,21	1,2,3,5,9,10,11,12,13,14,15,16,17,20
9	5, 19,3,21,7,17	1,2,4,6,8,10,11,12,13,14,15,16,18,20
10	3, 20,7,16,5,18	1,2,4,6,8,9,11,12,13,14,15,17,19,21
11	1, 21,4,18,2,20	3,5,6,7,8,9,10,12,13,14,15,16,17,19
12	4,17,1,20,6,15	2,3,5,7,8,9,10,11,13,14,16,18,19,21
13	6,14,5,15,1,19	2,3,4,7,8,9,10,11,12,16,17,18,20,21
14	6,13,2,17,3,16	1,4,5,7,8,9,10,11,12,15,18,19,20,21
15	2,16,5,13,6,12	1,3,4,7,8,9,10,11,14,17,18,19,20,21
16	2,15,7,10,3,14	1,4,5,6,8,9,11,12,13,17,18,19,20,21
17	4,12,2,14,7,9	1,3,5,6,8,10,11,13,15,16,18,19,20,21
18	7,8,4,11,5,10	1,2,3,6,9,12,13,14,15,16,17,19,20,21
19	5,9,6,8,1,13	2,3,4,7,10,11,12,14,15,16,17,18,20,21
20	3,10,1,12,2,11	4,5,6,7,8,9,13,14,15,16,17,18,19,21
21	1,11,3,9,4,8	2,5,6,7,10,12,13,14,15,16,17,18,19,20

Table 10.
Case 3 associates.

Case 4

$$t = 15, k = 3, r = 4, b = 20, \lambda_1 = 1, \lambda_2 = 0, n_1 = 8, n_2 = 6$$

$$P_1 = \begin{bmatrix} 4 & 3 \\ 3 & 3 \end{bmatrix}, P_2 = \begin{bmatrix} 7 & 1 \\ 1 & 4 \end{bmatrix}$$

$$E_1 = 0.725, E_2 = 0.6591, E = 0.6952$$

For **Case 1** an efficiency of E = 0 implies a 100% efficient design. By achieving E1 = E2 = E = 0, the allocations defined in the P1 and P2 matrices result in a maximally efficient balanced incomplete block design for estimating the main effects of the 3 factors over the 15 runs. **Case 2** defines a constant block-size with constant sum-block partially balanced incomplete block (PBIB) design. This achieves variances and overall design efficiency: E1 = 0.75 E2 = 0.6 E = 0.6364. An efficiency of 0.6364 implies the partial blocking and subsampling still allows reasonably efficient estimation of the main effects for the 3 factors at strength 2. The allocations balance runs across subsets to optimize estimation of 2-factor interactions at the cost of some design efficiency compared to a fully balanced design (E = 0).

				Sum
b1	1	11	21	33
b2	2	15	16	33
b3	3	10	20	33
b4	4	12	17	33
b5	5	9	19	33
b6	6	13	14	33
b7	7	8	18	33
b8	1	12	20	33
b9	2	14	17	33
b10	3	9	21	33
b11	4	11	18	33
b12	5	13	15	33
b13	6	8	19	33
b14	7	10	16	33
b15	1	13	19	33
b16	2	11	20	33
b17	3	14	16	33
b18	4	8	21	33
b19	5	10	18	33
b20	6	12	15	33
b21	7	9	17	33
b22	1	14	18	33
b23	2	12	19	33
b24	3	13	17	33
b25	4	9	20	33
b26	5	7	21	33
b27	6	11	16	33
b28	8	10	15	33

Table 11.
Case 4 design.

Case 3 also defines a constant block-size with constant sum-block partially balanced incomplete block (PBIB) design. This allocation achieves variances and efficiency: E1 = 0.7302, E2 = 0.6389 and E = 0.6751. The partial blocking balances runs across subsets to allow estimation of 3-factor interactions. An efficiency of 0.6751 indicates the subsample allocations still allow moderately efficient estimation of main effects, although less precisely than Case 1 or 2 constructions. The matrices P1 and P2 systematically distribute runs to balance representation across subsets at the cost of some variance inflation. In **Case 4**, the allocation of runs to the two subsamples allows

T	1st	2nd
1	11, 21,12,20,13,19,14,18	2,3,4,5,6,7,8,9,10,15,16,17
2	15,16,14,17,11,20,12,19	1,3,4,5,6,7,8,9,10,13,18,21
3	10, 20,9,21,14,16,13,17	1,2,4,5,6,7,8,11,13,15,18,19
4	12, 17,11,18,8,21,9,20	1,2,3,5,6,7,10,13,14,15,16,19
5	9, 19,13,15,10,18,7,21	1,2,3,4,6,8,11,12,14,16,17,20
6	13, 14,8,19,12,15,11,16	1,2,3,4,5,7,9,10,17,18,20,21
7	8, 18,10,16,9,17,5,21	1,2,3,4,6,11,12,13,14,15,19,20
8	7, 18,6,19,4,21,10,15	1,2,3,5,9,11,12,13,14,16,17,20
9	5, 19,3,21,7,17,4,20	1,2,6,8,10,11,12,13,14,15,16,18
10	3, 20,7,16,5,18,8,15	1,2,4,6,9,11,12,13,14,17,19,21
11	1, 21,4,18,2,20,6,16	3,5,7,8,9,10,12,13,14,15,17,19
12	4,17,1,20,6,15,2,19	3,5,7,8,9,10,11,13,14,16,18,21
13	6,14,5,15,1,19,3,17	2,4,7,8,9,10,11,12,16,18,20,21
14	6,13,2,17,3,16,1,18	4,5,7,8,9,10,11,12,15,19,20,21
15	2,16,5,13,6,12,8,10	1,3,4,7,9,11,14,17,18,19,20,21
16	2,15,7,10,3,14,6,11	1,4,5,8,9,12,13,17,18,19,20,21
17	4,12,2,14,7,9,3,13	1,5,6,8,10,11,15,16,18,19,20,21
18	7,8,4,11,5,10,1,14	2,3,6,9,12,13,15,16,17,19,20,21
19	5,9,6,8,1,13,2,12	3,4,7,10,11,14,15,16,17,18,20,21
20	3,10,1,12,2,11,4,9	5,6,7,8,13,14,15,16,17,18,19,21
21	1,11,3,9,4,8,5,7	2,6,10,12,13,14,15,16,17,18,19,20

Table 12.
Case 4 associates.

estimation of 3-factor interactions between the variables. The partial blocking distributes runs systematically though unequally across subsets. This achieves reasonable variance for assessing main effects, traded off against balance. Appropriate analysis can still estimate factor impacts. The matrices P1 and P2 define the subdivision rules that split the 15 runs to optimize estimation of 3-way interactions between the 3 factors. The efficiency factor, E, increases as the number of case increases (**Figures** 2 and 3).

2.3.3 Construction of design 3

The design t = 27, k = 3, λ_2 = 0 will be constructed in five cases since r = 7.
Case 1

$$t = 27, k = 3, r = 1, b = 9, \lambda_1 = 1, \lambda_2 = 0, n_1 = 2, n_2 = 24$$

$$P_1 = \begin{bmatrix} 1 & 0 \\ 0 & 24 \end{bmatrix}, P_2 = \begin{bmatrix} 0 & 2 \\ 2 & 21 \end{bmatrix}$$

$$E_1 = 0, E_2 = 0, E_3 = 0$$

				Sum
b1	1	11	21	33
b2	2	15	16	33
b3	3	10	20	33
b4	4	12	17	33
b5	5	9	19	33
b6	6	13	14	33
b7	7	8	18	33
b8	1	12	20	33
b9	2	14	17	33
b10	3	9	21	33
b11	4	11	18	33
b12	5	13	15	33
b13	6	8	19	33
b14	7	10	16	33
b15	1	13	19	33
b16	2	11	20	33
b17	3	14	16	33
b18	4	8	21	33
b19	5	10	18	33
b20	6	12	15	33
b21	7	9	17	33
b22	1	14	18	33
b23	2	12	19	33
b24	3	13	17	33
b25	4	9	20	33
b26	5	7	21	33
b27	6	11	16	33
b28	8	10	15	33
b29	1	15	17	33
b30	2	10	21	33
b31	3	11	19	33
b32	4	13	16	33
b33	5	8	20	33
b34	6	9	18	33
b35	7	12	14	33

Table 13.
Case 5 design.

T	1st	2nd
1	11, 21,12,20,13,19,14,18,15,17	2,3,4,5,6,7,8,9,10,16
2	15,16,14,17,11,20,12,19,10,21	1,3,4,5,6,7,8,9,13,18
3	10, 20,9,21,14,16,13,17,11,19	1,2,4,5,6,7,8,13,15,18
4	12, 17,11,18,8,21,9,20,13,16	1,2,3,5,6,7,10,14,15,19
5	9, 19,13,15,10,18,7,21,8,20	1,2,3,4,6,11,12,14,16,17
6	13, 14,8,19,12,15,11,16,9,18	1,2,3,4,5,7,10,17,20,21
7	8, 18,10,16,9,17,5,21,12,14	1,2,3,4,6,11,13,15,19,20
8	7, 18,6,19,4,21,10,15,5,20	1,2,3,9,11,12,13,14,16,17
9	5, 19,3,21,7,17,4,20,6,18	1,2,8,10,11,12,13,14,15,16
10	3, 20,7,16,5,18,8,15,2,21	1,4,6,9,11,12,13,14,17,19
11	1, 21,4,18,2,20,6,16,3,19	5,7,8,9,10,12,13,14,15,17
12	4,17,1,20,6,15,2,19,7,14	3,5,8,9,10,11,13,16,18,21
13	6,14,5,15,1,19,3,17,4,16	2,7,8,9,10,11,12,18,20,21
14	6,13,2,17,3,16,1,18,7,12	4,5,8,9,10,11,15,19,20,21
15	2,16,5,13,6,12,8,10,1,17	3,4,7,9,11,14,18,19,20,21
16	2,15,7,10,3,14,6,11,4,13	1,5,8,9,12,17,18,19,20,21
17	4,12,2,14,7,9,3,13,1,15	5,6,8,10,11,16,18,19,20,21
18	7,8,4,11,5,10,1,14,6,9	2,3,12,13,15,16,17,19,20,21
19	5,9,6,8,1,13,2,12,3,11	4,7,10,14,15,16,17,18,20,21
20	3,10,1,12,2,11,4,9,5,8	6,7,13,14,15,16,17,18,19,21
21	1,11,3,9,4,8,5,7,2,10	6,12,13,14,15,16,17,18,19,20

Table 14.
Case 5 associates.

Case 2

$$t = 27, k = 3, r = 2, b = 18, \lambda_1 = 1, \lambda_2 = 0, n_1 = 4, n_2 = 22$$

$$P_1 = \begin{bmatrix} 1 & 2 \\ 2 & 20 \end{bmatrix}, P_2 = \begin{bmatrix} 3 & 1 \\ 1 & 20 \end{bmatrix}$$

$$E_1 = 0.7667, E_2 = 0.6389, E_3 = 0.6557$$

Case 3

$$t = 27, k = 3, r = 3, b = 27, \lambda_1 = 1, \lambda_2 = 0, n_1 = 6, n_2 = 20$$

$$P_1 = \begin{bmatrix} 1 & 4 \\ 4 & 16 \end{bmatrix}, P_2 = \begin{bmatrix} 5 & 1 \\ 1 & 18 \end{bmatrix}$$

$$E_1 = 0.7284, E_2 = 0.6556, E_3 = 0.671$$

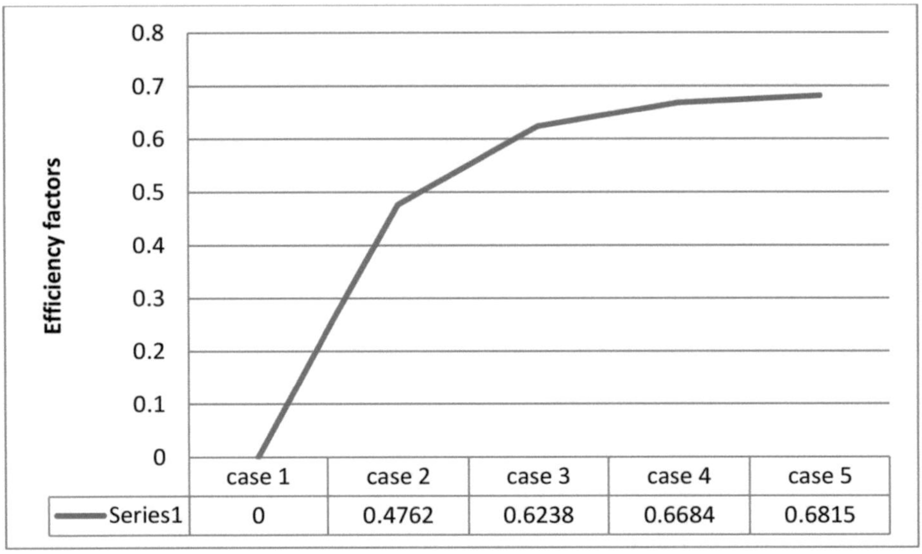

Figure 1.
Design parameters (t = 21, k = 3).

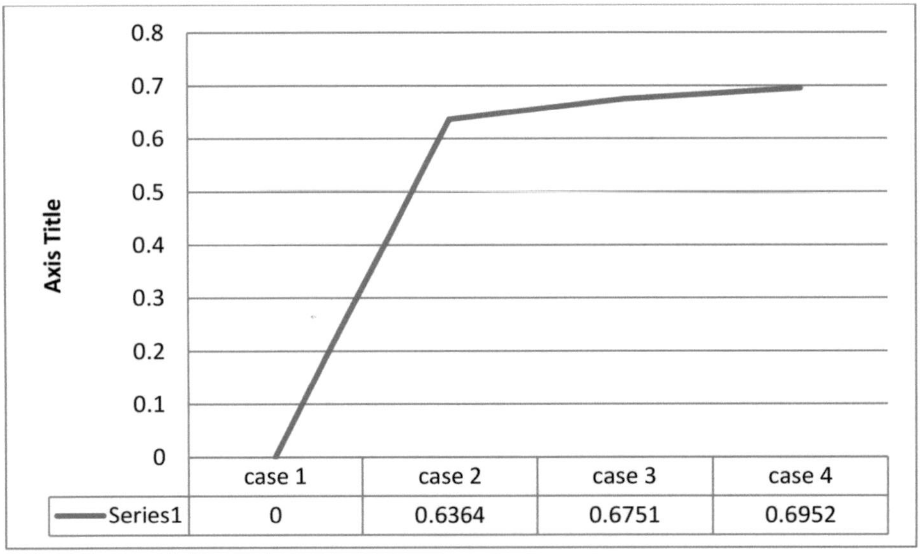

Figure 2.
Design parameters (t = 15, k = 3).

Case 4

$$t = 27, k = 3, r = 4, b = 36, \lambda_1 = 1, \lambda_2 = 0, n_1 = 8, n_2 = 18$$

$$P_1 = \begin{bmatrix} 1 & 6 \\ 6 & 12 \end{bmatrix}, P_2 = \begin{bmatrix} 7 & 1 \\ 1 & 16 \end{bmatrix}$$

$$E_1 = 0.7115, E_2 = 0.6607, E_3 = 0.6756$$

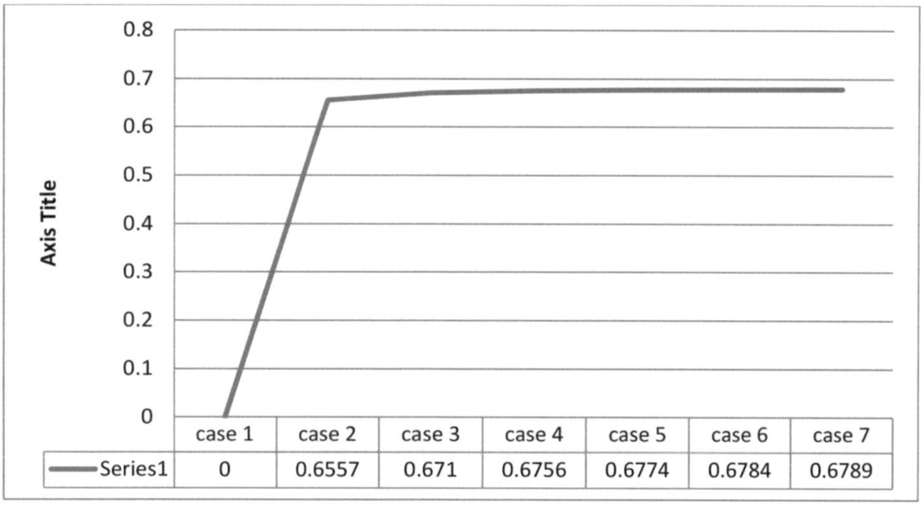

Figure 3.
Design parameters (t = 27, k = 3).

Case 5

$$t = 27, k = 3, r = 5, b = 45, \lambda_1 = 1, \lambda_2 = 0, n_1 = 10, n_2 = 16$$

$$P_1 = \begin{bmatrix} 1 & 8 \\ 8 & 8 \end{bmatrix}, P_2 = \begin{bmatrix} 9 & 1 \\ 1 & 14 \end{bmatrix}$$

$$E_1 = 0.702, E_2 = 0.663, E_3 = 0.6774$$

Case 6

$$t = 27, k = 3, r = 6, b = 54, \lambda_1 = 1, \lambda_2 = 0, n_1 = 12, n_2 = 14$$

$$P_1 = \begin{bmatrix} 1 & 10 \\ 10 & 4 \end{bmatrix}, P_2 = \begin{bmatrix} 11 & 1 \\ 1 & 12 \end{bmatrix}$$

$$E_1 = 0.6958, E_2 = 0.6641, E_3 = 0.6784$$

Case 7

$$t = 27, k = 3, r = 7, b = 63, \lambda_1 = 1, \lambda_2 = 0, n_1 = 14, n_2 = 12$$

$$P_1 = \begin{bmatrix} 1 & 11 \\ 11 & 2 \end{bmatrix}, P_2 = \begin{bmatrix} 13 & 1 \\ 1 & 10 \end{bmatrix}$$

$$E_1 = 0.6914, E_2 = 0.6648, E_3 = 0.6789$$

3. Data analysis

3.1 Dataset description

This dataset was collected on the impact of Body Mass Index (BMI) on impact oscillometry (IOS) measures on children with sickle cell diseases. The source of the dataset [26].

For the purpose of this chapter, 45 observations were extracted which has the following factors; asthma, hydroxyurea, ICS and LABA. Furthermore, each treatment was replicated 3 times alongside with the gender. The BMI results were recorded.

Table 15 dataset will be addressed in two cases (2 and 3) based on the priorities. The priority is based on the fixed sum of 24 (**Tables 16–21**).

SN	Subject ID	Hydroxyurea	Asthma	ICS	LABA	REP	Gender	BMI
1	1	Yes	Yes	Yes	Yes	1	Male	23.6
2	1	Yes	Yes	Yes	Yes	1	Male	23.74
3	8	Yes	Yes	Yes	Yes	1	Male	18.39
4	1	Yes	Yes	Yes	No	2	Male	22.84
5	1	Yes	Yes	Yes	No	2	Male	23.53
6	3	Yes	Yes	Yes	No	2	Male	20.95
7	3	Yes	Yes	No	No	3	Male	20.79
8	21	Yes	Yes	No	No	3	Male	15.87
9	30	Yes	Yes	No	No	3	Male	13.98
10	13	Yes	No	No	No	4	Male	16.36
11	13	Yes	No	No	No	4	Male	17.52
12	13	Yes	No	No	No	4	Male	19.07
13	25	No	Yes	Yes	Yes	5	Male	17.68
14	59	No	Yes	Yes	Yes	5	male	16.71
15	64	No	Yes	Yes	Yes	5	male	16.01
16	24	No	Yes	Yes	No	6	Male	15.36
17	25	No	Yes	Yes	No	6	Male	16.03
18	25	No	Yes	Yes	No	6	Male	16.26
19	24	No	Yes	No	No	7	Male	15.25
20	37	No	Yes	No	No	7	Male	15.09
21	37	No	Yes	No	No	7	Male	15.83
22	15	No	No	No	No	8	Male	22.03
23	36	No	No	No	No	8	Male	14.81
24	36	No	No	No	No	8	Male	14.83
25	10	Yes	Yes	Yes	Yes	9	Female	20.19
26	10	Yes	Yes	Yes	Yes	9	Female	20.49
27	10	Yes	Yes	Yes	Yes	9	Female	21.66
28	14	Yes	Yes	No	No	10	Female	16.02
29	14	Yes	Yes	No	No	10	Female	16.79
30	14	Yes	Yes	No	No	10	Female	17.8
31	57	No	Yes	Yes	Yes	11	Female	17.68
32	57	No	Yes	Yes	Yes	11	Female	16.82
33	57	No	Yes	Yes	Yes	11	Female	16.75

SN	Subject ID	Hydroxyurea	Asthma	ICS	LABA	REP	Gender	BMI
34	2	No	No	No	No	12	Female	27.82
35	4	No	No	No	No	12	Female	22.29
36	4	No	No	No	No	12	Female	22.35
37	12	Yes	Yes	Yes	No	13	Female	16.41
38	12	Yes	Yes	Yes	No	13	Female	17.39
39	12	Yes	Yes	Yes	No	13	Female	18.16
40	11	Yes	No	No	No	14	Female	16.13
41	11	Yes	No	No	No	14	Female	17.26
42	11	Yes	No	No	No	14	Female	21
43	7	No	Yes	Yes	No	15	Female	20.98
44	16	No	Yes	Yes	No	15	Female	20.8
45	44	No	Yes	Yes	No	15	Female	14.93

Table 15.
Table showing the impact of BMI on IOS.

				Sum
b1	1(23.6)	8(22.03)	15(20.98)	24
b2	1(23.74)	11(17.68)	12(27.82)	24
b3	2(22.84)	10(16.02)	12(22.29)	24
b4	2(23.53)	7(15.25)	15(20.8)	24
b5	3(20.79)	7(15.09)	14(16.13)	24
b6	315.87)	8(14.81)	13(16.41)	24
b7	4(16.36)	9(20.19)	11(16.82)	24
b8	4(17.52)	6(15.36)	14(17.26)	24
b9	5(17.68)	6(16.03)	13(17.39)	24
b10	5(16.71)	9(20.49)	10(16.79)	24

Table 16.
The dataset arranged in blocks when r = 2.

Case 1
When r = 2

$$t = 15, k = 3, r = 2, b = 10$$

$$\text{Let } p = \frac{rG}{N} = \frac{2\,x\,564.28}{30} = 37.6187$$

Then $q_1 = 45.283–37.62 = 7.66$

$$Q_1 = 47.34–45.283 = 2.06$$

$$SSB = (3)–(1)$$

Block				B	$\frac{B}{k}$	$\frac{B^2}{k}$	B^2
b1	1	8	15				
	23.6	22.03	20.98	66.61	22.2	1479	4436.89
b2	1	11	12				
	23.74	17.68	27.82	69.24	23.08	1598.1	4794.18
b3	2	10	12				
	22.84	16.02	22.29	61.15	20.38	1246.4	3739.32
b4	2	7	15				
	23.53	15.25	20.8	59.58	19.86	1183.3	3549.78
b5	3	7	14				
	20.79	15.09	16.13	52.01	17.34	901.68	2705.04
b6	3	8	13				
	15.87	14.81	16.41	47.09	15.7	739.16	2217.47
b7	4	9	11				
	16.36	20.19	16.82	53.37	17.79	949.45	2848.36
b8	4	6	14				
	17.52	15.36	17.26	50.14	16.71	838.01	2514.02
b9	5	6	13				
	17.68	16.03	17.39	51.1	17.03	870.4	2611.21
b10	5	9	10				
	16.71	20.49	16.79	53.99	18	971.64	2914.92
Total				564.3		10,777	32331.2

Table 17.
Block estimation.

$$SS_{Total} = (2)–(1)$$

The model assumption

$$Y_{ij} = \mu + \alpha_i + \beta_j + e_{ij}$$

Where α is the treatment, β is the block, e_{ij} is the error term and μ is the mean.

The Anova **Table 22** shows that both the blocks and the treatments adjustment are significant (p-value ˂0.05). Therefore, we can identify the treatments responsible for the significance (**Figure 4**).

The **Table 23** shows that treatments 1,2,12 and 15 contributed to the significant of the results (**Table 24**)

Case 2

When r = 3, then

$$t = 15, k = 3, r = 3 \text{ and } b = 15$$

T	b1	b2	b3	b4	b5	b6	b7	b8	b9	b10	T	Q	q
1	23.6	23.74									47.34		
	22.2	23.08									45.283	2.06	7.66
2			22.84	23.53							46.37		
			20.38	19.86							40.243	6.13	2.62
3					20.79	15.87					36.66		
					17.34	15.7					33.033	3.63	−4.6
4							16.36	17.52			33.88		
							17.79	16.71			34.503	−0.62	−3.1
5									17.68	16.71	34.39		
									17.03	18	35.03	−0.64	−2.6
6								15.36	16.03		31.39		
								16.71	17.03		33.747	−2.36	−3.9
7				15.25	15.09						30.34		
				19.86	17.34						37.197	−6.86	−0.4
8	22.03					14.81					36.84		
	22.2					15.7					37.9	−1.06	0.28
9							20.19			20.49	40.68		
							17.79			18	35.787	4.89	−1.8
10			16.02							16.79	32.81		
			20.38							18	38.38	−5.57	0.76
11		17.68					16.82				34.5		
		23.08					17.79				40.87	−6.37	3.25
12		27.82	22.29								50.11		
		23.08	20.38								43.463	6.65	5.84
13						16.41			17.39		33.8		
						15.7			17.03		32.73	1.07	−4.9
14					16.13			17.26			33.39		
					17.34			16.71			34.05	−0.66	−3.6
15	20.98			20.8							41.78		
	22.2			19.86							42.063	−0.28	4.44

Table 18.
Treatments estimation.

Following the similar calculation, the Anova table was estimate in **Table 25**

Table 25 showed that both the blocks and the treatments are not significant since the p-value is greater than 0.05

G	1	2	3	G_i
1	1	6	11	
	2.06	−2.36	−6.37	−6.67
2	2	7	12	
	6.13	−6.86	6.65	5.92
3	3	8	13	
	3.63	−1.06	1.07	3.64
4	4	9	14	
	−0.62	4.89	−0.66	3.61
5	5	10	15	
	−0.64	−5.57	−0.28	−6.49

Table 19.
Group for treatment adjustment.

T	Q_i	G_i	$W = \frac{k}{r(k-1)}Q_i$	$Z = \frac{1}{r(k-1)}G_i$	T = W-Z	TQ_i
1	2.06	−6.67	1.545	−1.6675	3.2125	6.61775
2	6.13	5.92	4.5975	1.48	3.1175	19.11028
3	3.63	3.64	2.7225	0.91	1.8125	6.579375
4	−0.62	3.61	−0.465	0.9025	−1.3675	0.84785
5	−0.64	−6.49	−0.48	−1.6225	1.1425	−0.7312
6	−2.36	−6.67	−1.77	−1.6675	−0.1025	0.2419
7	−6.86	5.92	−5.145	1.48	−6.625	45.4475
8	−1.06	3.64	−0.795	0.91	−1.705	1.8073
9	4.89	3.61	3.6675	0.9025	2.765	13.52085
10	−5.57	−6.49	−4.1775	−1.6225	−2.555	14.23135
11	−6.37	−6.67	−4.7775	−1.6675	−3.11	19.8107
12	6.65	5.92	4.9875	1.48	3.5075	23.32488
13	1.07	3.64	0.8025	0.91	−0.1075	−0.11503
14	−0.66	3.61	−0.495	0.9025	−1.3975	0.92235
15	−0.28	−6.49	−0.21	−1.6225	1.4125	−0.3955
Total						151.2204

Table 20.
Treatment adjustment estimates.

4. Discussion

This study has introduced Constant Block-Size with Constant Sum-Block Partially Balanced Incomplete Block Designs (CBS-CSB PBIBDs) as a new class of partially balanced incomplete block designs that will be used when there is need to prioritize the treatments combination. This study defined constant block-size; a design is said to

(1)	$\frac{G^2}{N} = \frac{564.3 \times 564.3}{30}$	**10614.483**
(2)	$\sum x^2$	10939.08
(3)	$\frac{\sum B^2}{k}$	10,777

Table 21.
Sum of square table.

SV	SS	df	MS	F	P-value
blocks	162.517	9	18.0574	9.977	0.00557***
Treatments adj	151.2204	14	10.8015	5.968	0.0186***
Error	10.8596	6	1.8099		
Total	324.597	29			

*Note: Estimates with "****" means they are significant since the p < 0.05 is considered significant in the study.*

Table 22.
ANOVA for the Design when r = 2.

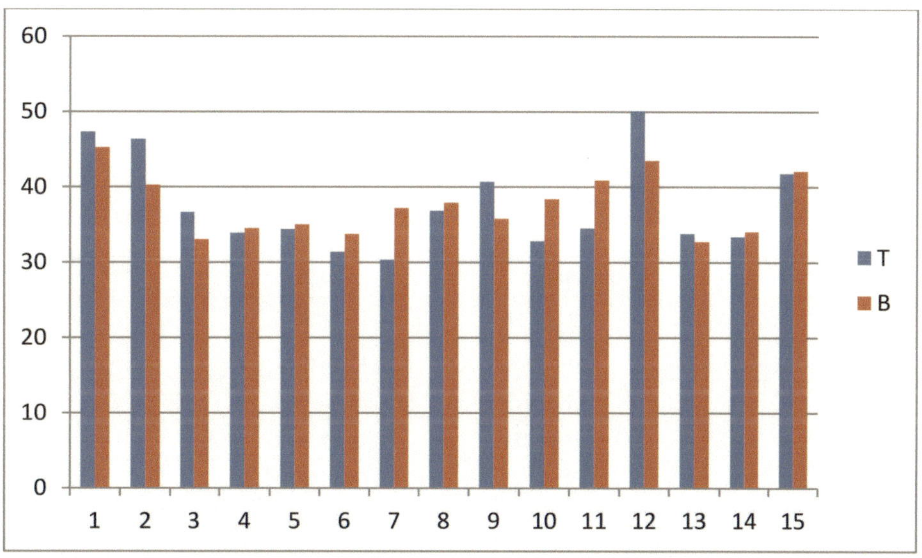

Figure 4.
Chart showing the treatment and block means.

have a constant block-size if the number of experimental unit, k, are the same irrespective of the number of blocks, b. Also the study defined constant block-sum; a design is said to be constant block-sum if the number of blocks, b, are the same, $b = \frac{t}{k}$, and the sum of each corresponding number of experimental units are the same, irrespective of the number of treatments and replicates [24]. It is determined by the sum of all the treatments divided by number of blocks, $CBS = \frac{\sum t}{b}$. Furthermore, this study gave conditions on constant sum-block; there is constant block size, each corresponding blocks has constant sum, the number of blocks are not the same, $b = \frac{tr}{k}$,

T	T_i	B_i	Mean
1	47.34	45.283	23.67
2	46.37	40.243	23.185
3	36.66	33.033	18.33
4	33.88	34.503	16.94
5	34.39	35.03	17.195
6	31.39	33.747	15.695
7	30.34	37.197	15.17
8	36.84	37.9	18.42
9	40.68	35.787	20.34
10	32.81	38.38	16.405
11	34.5	40.87	17.25
12	50.11	43.463	25.055
13	33.8	32.73	16.9
14	33.39	34.05	16.695
15	41.78	42.063	20.89

Table 23.
Treatment, block and mean.

				Sum
b1	1(23.6)	8(22.03)	15(20.98)	24
b2	1(23.74)	11(17.68)	12(27.82)	24
b3	1(18.39)	10(16.02)	13(16.41)	24
b4	2(22.84)	10(16.79)	12(22.29)	24
b5	2(23.53)	7(15.25)	15(20.8)	24
b6	2(20.95)	8(14.81)	14(16.13)	24
b7	3(20.79)	7(15.09)	14(17.26)	24
b8	3(15.87)	8(14.83)	13(17.39)	24
b9	3(13.98)	9(20.19)	12(22.35)	24
b10	4(16.36)	9(20.49)	11(16.82)	24
b11	4(17.52)	6(15.36)	14(21)	24
b12	4(19.07)	5(16.01)	15(14.93)	24
b13	5(17.68)	6(16.03)	13(18.16)	24
b14	5(16.71)	9(21.66)	10(17.8)	24
b15	6(16.26)	7(15.83)	11(16.82)	24

Table 24.
Table showing $r = 3$.

SV	SS	df	MS	F	P-value
Blocks	186.569	14	13.326	2.1896	0.0674
Treatments adj	150.09	14	10.721	1.7616	0.1385
Error	97.373	16	6.086		
Total	434.032	44			

Table 25.
ANOVA for the design when r = 3.

the number of treatments, t, are constant and the number of replicates, r, are at least one, $r \geq 1$. This study also gave conditions for Constant Block-Size with Constant Sum-Block Partially Balanced Incomplete Block Designs, CBS-CSB PBIBDs; the treatments combinations must be prioritized, that is, there must be number of zero associates, $\lambda_2 = 0$, the maximum number of replicates must be necessarily determine from treatment one and the number of treatments and block sizes remain fixed. The study designed t = 21, k = 3; t = 15, k = 3; t = 27, k = 3. And their respective cases were designed and presented with line charts. It was observed when the number of replicate, r, increases the efficiency factors increases. It was also observed that when r = 1, then E = 0.

After the design of the experiments, the study analyzes the experiments using real-life application dataset. The dataset was collected on the impact of Body Mass Index (BMI) on impact oscillometry (IOS) measures on children with sickle cell diseases. [26]. Forty-five (45) observations were extracted which has the following factors; asthma, hydroxyurea, ICS and LABA. Furthermore, each treatment was replicated 3 times alongside with gender. The BMI results were recorded. The dataset was addressed in two cases (2 and 3) based on the priorities. The priority is based on the fixed sum of 24. For the case one, r = 2, the study shows that both the blocks and the treatments adjustment are significant since the p-value is less than 0.05. But the case two, r = 3, the study shows that both the blocks and the treatments are not significant since the p-value is greater than 0.05.

Consequently, the study is not applicable on the even number of treatments, i.e., when t = 2, 4, 6 ... Even number of treatments will never give a constant block size [15]. In that case, there is need to research on constant sum-block with even number of treatments.

5. Conclusion

In conclusion, Constant Block-Size with Constant Sum-Block Partially Balanced Incomplete Block Designs (CBS-CSB PBIBDs) has been constructed in this study as a new class of partially balanced incomplete block designs that should be used when it requires prioritizing the treatments combination. The CBS-CSB PBIBDs are designed to provide efficient and precise estimates of treatment effects and they are useful in obtaining maximum information from the available resources, leading to increased experimental efficiency. The CBS-CSB PBIBDs constructed are applicable for testing the impact of various factors on product quality, manufacturing processes, or other parameters of interest. Constant sum-block with even number of treatments will be considered for future research.

Author details

Babatunde L. Adeleke[1], Gabriel O. Adebayo[1] and Kazeem A. Osuolale[2*]

1 Faculty of Physical Science, Department of Statistics, University of Ilorin, Kwara State, Nigeria

2 Biostatistics Unit, Nigerian Institute of Medical Research, Yaba, Lagos State, Nigeria

*Address all correspondence to: whereisqosimadewale@gmail.com

References

[1] Bose RC, Nair KR. Partially balanced incomplete block designs. Sankhyā: The Indian Journal of Statistics. 1939;**4**: 337-372

[2] Bose RC, Shimamoto T. Classification and analysis of partially balanced incomplete block designs with two associate classes. Journal of the American Statistical. 1952;**47**:151-184

[3] Nair KR. Analysis of partially balanced incomplete block designs illustrated on the simple square and rectangular lattices. Biometrics. 1952; **8**(2):122-155

[4] Clatworthy WH. Partially balanced incomplete block designs with two associate classes and two treatments per block. Journal of Research of the National Bureau of Standards. 1955; **54**(2):177-190

[5] Hinkelmann K. Extended group divisible partially balanced incomplete block designs. The Annals of Mathematical Statistics. 1964;**35**(2): 681-695

[6] Mesner DM. A new family of partially balanced incomplete block designs with some Latin square design properties. The Annals of Mathematical Statistics. 1967; **38**(2):571-581

[7] Clatworthy WH, Cameron JM, Speckman JA. Tables of two-associate class partially balanced designs. In: NBS Applied Mathematics Series 63. Washington, DC: National Bureau of Standards; 1973

[8] Garg DK. Pseudo new modified Latin Square (NML m (m)) type PBIB designs. Communications in Statistics— Theory and Methods. 2010;**39**(19): 3485-3491

[9] Kaur P, Garg DK. Construction of some higher associate class PBIB designs using symmetrically repeated differences. Arya Bhatta Journal of Mathematics and Informatics. 2016;**8**(2): 65-78

[10] Sharma K, Garg DK. Construction of three associate PBIB designs using some sets of initial blocks. International Journal of Agricultural and Statistical Sciences. 2017;**13**(1):55-60

[11] Garg DK, Singh GP. Construction of four associate class PBIB designs using pairing in triplets system. Advances and Applications in Statistics. 2014;**41**(1):11

[12] Garg DK, Farooq SA. Construction of PBIB designs through chosen lines and triangles of graphs. International Journal of Mathematics Trends and Technology. 2014;**8**(1):25-32

[13] Saurabh S, Singh MK. Construction of some series of PBIB designs. Journal of Computer and Mathematical Sciences. 2018;**9**(12):2031-2036

[14] Singh GP, Garg DK. Construction of PBIB designs by using Boolean algebra. Journal of Computer and Mathematical Sciences. 2018;**9**(2):56-62

[15] Khattree R. A note on the nonexistence of the constant block-sum balanced incomplete block designs. Communications in Statistics-Theory and Methods. 2019;**48**(20):5165-5168

[16] Khattree R. The Parshvanath yantram yields a constant-sum partially balanced incomplete block design. Communications in Statistics-Theory and Methods. 2019;**48**(4):841-849

[17] Khattree R. On construction of constant block-sum partially balanced

incomplete block designs. Communications in Statistics-Theory and Methods. 2020;**49**(11):2585-2606

[18] Kaur P, Sharma K. Constant block-sum Youden-m square and PBIB designs using galois field. Communications in Some - Simulation and Computation. 2023;**52**(11):5396-5400

[19] Bansal N, Garg DK. Construction of some new Youden-m squares, incomplete Sudoku squares and PBIB designs. Communications in Statistics-Simulation and Computation. 2022:1-15. DOI: 10.1080/03610918.2022.2109674

[20] Gupta S. Constant block-sum designs through confounded factorials. Statistics and Applications. 2021;**20**: 93-101

[21] Gupta S. Constant block-sum two-associate class group divisible designs. Statistics and Applications. 2021;**19**(1): 141-148

[22] Khattree R. On construction of equireplicated constant block-sum designs. Communications in Statistics-Theory and Methods. 2022;**51**(13): 4434-4450

[23] Bansal N, Garg DK, Khattree R. Some results on the existence or nonexistence of constant block-sum triangular and other related designs. Journal of Statistical Theory and Practice. 2023;**17**(1):1

[24] Khattree R. Constant and nearly constant block-sum partially balanced incomplete block designs and magic rectangles. Journal of Statistical Theory and Practice. 2023;**17**(1):8

[25] Kaur P, Sharma K, Garg DK, Sharma MK. Package 'PBIBD'. Comprehensive R Archive Network (CRAN): O'Reilly Media, Inc.; 2017

[26] Utkarsh Singh. Impact of BMI on IOS measures on children with sickle cell diseases. 2023. Available from: Kaggle.com

Chapter 6

Perspective Chapter: Experimental Analysis of Black Hole Algorithm with Heuristic Algorithms in Traveling Salesman Problem

Mehmet Fatih Demiral

Abstract

Black hole algorithm (BHA) is a popular metaheuristic algorithm proposed and applied for data clustering in 2013. BHA was applied to continuous and discrete problems; it is also hybridized with some algorithms in the literature. The pure BHA shows better performance than others in discrete optimization, such as traveling salesman problems. However, it requires improving the algorithm with competitive heuristics. Many heuristics have often been used to construct the initial tour of a salesman, such as the nearest neighbor algorithm (NN), nearest insertion algorithm (NI), cheapest insertion algorithm (CI), random insertion algorithm (RI), furthest insertion algorithm (FI), and minimal spanning tree algorithm (MST). In addition, the black hole algorithm is combined with popular heuristics, such as swap/or insert, reverse/or 2-opt swap, and swap-reverse/or 3-opt swap, and tested with proper parameters in this study. In the experimentation, classical datasets are used via TSP-library. The experimental results are given as best, average solutions/or deviations, and CPU time for all datasets. Besides, the hybrid algorithms demonstrate a better performance rate to get optimality. Finally, hybrid algorithms solve the discrete optimization problem in a short computing time for all datasets.

Keywords: black hole algorithm, combinatorial optimization, heuristics, metaheuristics, traveling salesman problem

1. Introduction

Meta-heuristic algorithms are nature-inspired heuristics that are the subject of operations research, computer science, industrial engineering, transportation science, and other research fields [1–3]. The algorithms are generally applied to continuous and global optimization problems. Most algorithms are successful at continuous space, but others give average performance on continuous problems. On the other hand, it is needed to improve and develop variations of metaheuristics to handle with discrete problems in discrete space [4, 5]. Many heuristics have often been used to construct a traveling tour of a salesman, such as initial algorithms; nearest neighbor algorithm (NN), nearest insertion algorithm (NI), cheapest insertion algorithm (CI), random insertion algorithm (RI), furthest insertion algorithm (FI), minimal spanning tree algorithm (MST), and Christofides' heuristic algorithm (CH) [6]. Other heuristics are used to obtain improved new solutions, such as tour improvement algorithms; 2-opt, 3-opt, and k-opt algorithms proposed by Lin and Kernighan [7–9]. Mixed methods are the alternative to implement on constructed tours by 2-opt and 3-opt algorithms. Some tours may be raised from the optimal solution of a traveling salesman, so it is needed to eliminate them from the solution to obtain a complete tour. There should be added some constraints to the available linear programming model for traveling salesman in the cutting plane algorithm (CP). In the branch and bound method (BB), the problem is separated into different sub-problems. Then, this process continues to examine all sub-problems. The branch and cut algorithm (BC) is similar to the branch and bound algorithm. However, the local bounds are calculated via the cutting plane algorithm [10–13].

Modern heuristic methods have been investigated to find optimal, near-optimal, and acceptable solutions to combinatorial problems in recent decades. Classical metaheuristics have often been used to solve continuous and engineering problems, such as simulated annealing (SA), tabu search (TS), ant colony optimization (ACO), particle swarm optimization (PSO), and genetic algorithm (GA) [14, 15]. Modern metaheuristics have also been used to solve those problems, such as artificial bee colony algorithm (ABC), harmony search algorithm (HS), artificial atom algorithm (A3), black hole algorithm (BH), worm optimization (WO), water-wave optimization (WWO), camel algorithm (CA), whale optimization algorithm (WOA), socio evolution and learning optimization (SELO), physarum-energy optimization (PEO), color harmony algorithm (CHA), stochastic paint optimizer (SPO), and fire hawk optimizer (FHO) [16–20].

Population-based meta-heuristics generally search for optimal solutions better than single or multisolution-based algorithms. The reason for that is that they search for a wide area of solution space and focus on optimal areas by escaping local solutions. However, single or multisolution-based algorithms run quite better at long iterations and CPU time, such as hill climbing (HC), simulated annealing, and tabu search [15]. While population-based algorithms find near-optimal or acceptable solutions to optimization problems in short iterations and CPU time, they often solve problems optimally in long iterations and CPU time [16, 17]. Although the optimization process has frequently been done by iteration, most of the comparisons are done by population size, CPU time, the percentage of best and average deviations, and the number of objective function evaluations [21, 22].

The rest of the paper is organized as follows: In Section 2, the traveling salesman problem is clearly explained. The black hole algorithm, its variations, and the pseudocode are given in Section 3. In Section 4, the experimental analysis of the application of the proposed algorithms is described, and finally, Section 5 includes a conclusion and future works.

2. Traveling salesman problem

The traveling salesman problem is a wide research area that has been worked on by many scientists. TSP is a combinatorial problem, yet its optimal solution has not been found in polynomial time. Many heuristic algorithms have been proposed for combinatorial problems, but none of them guarantees the optimal solution in a reasonable time [23]. There exist many variants of TSP, such as symmetric (s-TSP) and asymmetric traveling salesman problem (a-TSP), traveling purchaser problem (TPP), Hamiltonian cycle problem (HCP), sequential ordering problem (SOP), multiple traveling salesman problem (m-TSP), orienteering problem (OP), vehicle routing problem (VRP), and its variants [24]. Vehicle routing problem has also important variants that are most applicable in transportation, industry, engineering, and science such as vehicle routing problem with profits (VRPP), team orienteering problem (TOP), capacitated team orienteering problem (CTOP), the top with time windows (TOPTW), vehicle routing problem with pick-up and delivery (VRPPD), vehicle routing problem with LIFO, vehicle routing problem with time windows (VRPTW), capacitated vehicle routing problem (CVRP or CVRPTW), open vehicle routing problem (OVRP), inventory routing problem (IRP), and vehicle routing problem with transfers (VRPWT) [25–27]. Its variants can be reduced to a traveling salesman problem if certain assumptions are accepted or any alternative algorithm is found. The problem gets hard when the number of data increases, the distance matrix changes, some constraints are added, the objective function changes, and so on.

The popular variants of TSP are traveling purchaser problems, multiple traveling salesman problems, and vehicle routing problems. The traveling purchaser problem is a problem that purchaser travels and purchases a set of items from different places if a budget, fuel, time, and other constraints are satisfied. In addition, different objectives can be set under the constraints [28, 29]. The multiple traveling salesman problem is a problem in which more than one salesman travels and turns to the home under different objectives [30]. The vehicle routing problem (VRP) is a combinatorial problem to find the optimal set of routes for a fleet of vehicles to traverse to deliver to a given set of customers. The fleet of vehicles can be set as homogeneous or heterogeneous [27].

The formulation of TSP is briefly given in Eq. (1). The nodes, edges, sets, and vertices are given as follows: N is the set of n nodes, E is the set of the edges, and $Dist_{ij} = (d_{ij})$ is the symmetric distances $(dij = dji)$ that are equal between nodes and the asymmetric distances $(d_{ij} \neq d_{ji})$ that are not equal between nodes.

$Tour_k = \{v_1, v_2, \dots \dots, v_n, v_1\}$ is a candidate tour for all the solution space, and k = 1,2,3, … …,m. v_1 indicates the first vertex, and v_n indicates the nth vertex of all the candidate tours.

$$Min. \sum_{i=1}^{n-1} \left(d_{v_i, v_{i+1}}\right) + d_{v_n, v_1} \tag{1}$$

Different TSP distances can be used in the solution of the discrete problem. Therefore, the optimal solution, deviation of solution, and CPU time change when different types of distances are used. The Euclidean distance is generally applied to calculate the distance between nodes using Eq. (2). In Eq. (2), x_i and x_j indicate the x-axis coordinates of Euclidean cities, and y_i and y_j indicate the y-axis coordinates of the Euclidean cities.

$$d_{i,j} = \sqrt{(x_i - x_j)^2 + \left(y_i - y_j\right)^2} \tag{2}$$

3. Black hole algorithm and its hybrids

The black hole algorithm is a metaheuristic algorithm proposed for data clustering in 2013. The algorithm principle is based on the black hole phenomenon and the stars that move toward the black hole. There is a possibility that a star reaches a better location in continuous space or a better objective in discrete space [31, 32]. Thus, the new star replaces the black hole. If a candidate star (solution) enters the radius of the event horizon in the algorithm, the candidate star will be sucked by the black hole. When a candidate star dies, another candidate star (solution) is randomly generated in the search space. Then, the old stars (solutions) and new stars create a new population [33–35].

The new candidates are generated in the continuous search space via Eq. (3).

$$x_i(t+1) = x_i(t) + rand * (x_{BH} - x_i(t)) \tag{3}$$

In this discrete application, update star locations via the selection of minimum multiple neighborhood operators in Eq. (4).

$$MMOp = \min(swap, insertion, reversing, swap - reversing) \tag{4}$$

The new candidates are updated by the objectives found in Eq. (5).

$$\begin{aligned} NewObj^{i,t}_{now} &< Obj^{i,t}_{now} \\ Obj^{i,t}_{now} &= NewObj^{i,t}_{now} \end{aligned} \tag{5}$$

The black hole algorithm event-horizon condition in continuous search space is formulated in Eq. (6).

$$|f_{BH} - f_i| < \frac{f_{BH}}{\sum_{i=1}^{N} f_i} \tag{6}$$

The event-horizon condition is applied, and the specific parameter Q_{Data} is selected for discrete problems especially in each TSP dataset via Eqs. (6) and (7) [36].

$$\frac{Q_{Data}}{|Obj_{BH} - Obj_i|} < \frac{f_{BH}}{\sum_{i=1}^{N} f_i} \tag{7}$$

When the star enters the event-horizon radius of the black hole, the old star (solution) is updated by the new solution. The final population is generated by the current and the updated solutions. The black hole algorithm and its hybrids have a pure algorithm structure and less number of parameters; so, their use is effective and advantageous for scientific areas, technology, and research.

The pseudo-code of the improved and/or hybrid black hole algorithm is shown in **Figure 1** [36].

The black hole algorithm can be hybridized with a constructive heuristic (Clarke and Wright, NN, k-NN, NI, CI, RI, FI, MST, or CH), with improving a local search (swap/or insert, reverse/or 2-opt swap, and swap-reverse/or 3-opt swap heuristic).

As noted above, even though those hybrid algorithms do not guarantee optimal solutions, they could solve the combinatorial problems in competitive CPU times. Most of the hybrid algorithms run in polynomial computational times, and they produce near-optimal or acceptable solutions, which are useful for science, engineering, and technology due to some criteria, such as security, design, flexibility, assembly, and production.

Proposed Algorithm: Hybrid Black-Hole Algorithm (BH+Heuristics)

Initialize the Population with a constructive
heuristic (Clarke and Wright (CW), NN, k-NN, NI, CI, RI, FI, MST, or CH)
Initialize Black Hole algorithm dataset parameter (Q)
Compute Population objective function
and determine the black hole
While (Counter < Maximum Iteration Number)
For i=1: Population of Candidate Stars
Update solutions using Eq. (4)
Apply an improvement heuristic (swap/or insert, reverse/or 2-opt swap,
swap-reverse/or 3-opt swap)
Exchange the solution with the black hole via the fitness function
End For i
Decide the acceptance of new solutions using Eq. (5)
If (event horizon condition is applied using Eq. (7))
Replace it with a random solution and create a new population
End If
Search the population and the best candidate (BC) as
the black hole (BH)
End While
State the Results (Best, Average, Worst, Std. Dev., and CPU time

Figure 1.
Pseudo-code of the improved and/or hybrid black hole algorithm.

4. Experimental results

The 10 small and medium-scale datasets (52–195) are used from the TSPLIB library in the experimental study. In the study, all the computations were run 10 times independently on Intel® Core™ i7 12,650-H CPU 2.3–4.7 GHz speed with 64 GB RAM using MATLAB. In **Table 1**, the performance of hybrid metaheuristic algorithms BH + swap, BH + insertion, BH + reverse, and BH + swap-reverse are evaluated by algorithm results and percentage deviation. In this work, a type of 2-opt swap heuristic named reverse and one type of 3-opt swap heuristic named swap-reverse heuristics are used. In **Table 1**, BH + NN + reverse and other BH hybrid algorithms are also compared with each other. The standard appropriate) parameters are used for all the datasets using 1000–3000 iteration numbers and an increasing number of data (52–195).

The population size of algorithms BH + Heuristics is increased two times and taken as 100 for (berlin52-rat99), and 200 for (kroa100-rat195) in **Table 1**. The maximum iteration number is increased three times and taken as 1000 for (berlin52-st70), 2000 for (eil76-rat99), and 3000 for (kroa100-rat195) in **Table 1**.

In **Table 1**, for berlin52-bier127, the hybrid metaheuristics without an initial algorithm would give challenging near-optimal solutions. Particularly, in kroa150 and rat195, the deviations from best-known solutions can be simply observed. On the other hand, the swap-reverse heuristic gives worse solutions than the 3-opt swap heuristic algorithm. Because, if a 3-opt heuristic was applied, the best among the eight alternatives was taken. As concluded from **Table 2**, though the BH + NN + reverse heuristic CPU times are slightly different than the BH + insertion heuristic times, the obvious differences can be observed in the Best and Average deviations. This is valid for other BH hybrid algorithms (BH + swap, BH + reverse, and BH + swap-reverse). The conclusion drawn from **Tables 1** and **2** is that the near-optimal and acceptable solutions are preferred by the decision-makers and managers who work with science, engineering, and technology due to some criteria mentioned, such as security, design, and flexibility.

Problem	Algorithm	Best	Ave.	BDev.	AvDev.	Time
berlin52	BH + swap	7769.84	7910.85	3.02	4.89	3.43
(7542)	BH + insertion	7635.3	7901.93	1.24	4.77	3.15
	BH + reverse	7583.71	7842.05	0.55	3.98	3.31
	BH + s-reverse	7899.49	8074.09	4.74	7.06	4.16
	BH + NN + reverse	7760.63	7785.24	2.9	3.23	3.71
st70	BH + swap	742.02	765.37	9.93	13.39	4.32
(675)	BH + insertion	709.7	756.13	5.14	12.02	4.1
	BH + reverse	708.84	743.11	5.01	10.09	3.67
	BH + s-reverse	772.66	794.77	14.47	17.74	4.88
	BH + NN + reverse	695.39	713.96	3.02	5.77	4.6
eil76	BH + swap	580.71	588.76	7.94	9.43	10.51
(538)	BH + insertion	578.7	585.95	7.56	8.91	10.45
	BH + reverse	575.46	583.98	6.96	8.55	9.4
	BH + s-reverse	566.9	583.39	5.37	8.44	13.09
	BH + NN + reverse	559.37	570.9	3.97	6.12	10.45
pr76	BH + swap	115,104	118,241	6.42	9.32	6.18
(108159)	BH + insertion	113,677	116,393	5.1	7.61	8.7
	BH + reverse	110,053	114,910	1.75	6.24	9.22
	BH + s-reverse	113,407	116,373	4.85	7.59	14.82
	BH + NN + reverse	110,919	112,468	2.55	3.98	9.89
rat99	BH + swap	1437.76	1474.48	18.72	21.76	9.3
(1211)	BH + insertion	1341.74	1453.95	10.8	20.06	9.87
	BH + reverse	1339.32	1399.75	10.6	15.59	12.16
	BH + s-reverse	1424.29	1479.18	17.61	22.15	12.6
	BH + NN + reverse	1278.04	1300.53	5.54	7.39	9.55
kroa100	BH + swap	22665.8	23788.2	6.5	11.78	26.1
(21282)	BH + insertion	22385.2	23697.5	5.18	11.35	24.06
	BH + reverse	22,550	23988.5	5.96	12.72	32.29
	BH + s-reverse	22465.1	23591.1	5.56	10.85	43.48
	BH + NN + reverse	21311.7	21920.1	0.14	3	28.93
eil101	BH + swap	677.92	702.8	7.78	11.73	30.72
(629)	BH + insertion	670.43	696.2	6.59	10.68	27.6
	BH + reverse	671.46	691.66	6.75	9.96	53.65
	BH + s-reverse	684.11	708.91	8.76	12.7	32.5
	BH + NN + reverse	655.14	662.11	4.16	5.26	32.12
bier127	BH + swap	131,090	137,349	10.83	16.12	42.1
(118282)	BH + insertion	127,183	134,630	7.53	13.82	35.9
	BH + reverse	127,443	135,921	7.74	14.91	39.69

Problem	Algorithm	Best	Ave.	BDev.	AvDev.	Time
	BH + s-reverse	131,046	136,960	10.79	15.79	31.93
	BH + NN + reverse	123,558	125,358	4.46	5.98	29.11
kroa150	BH + swap	33735.6	36374.7	27.19	37.14	48.3
(26524)	BH + insertion	33340.2	35639.8	25.7	34.37	40.51
	BH + reverse	33319.1	35,195	25.62	32.69	30.89
	BH + s-reverse	34169.1	36700.8	28.82	38.37	45.34
	BH + NN + reverse	28533.9	28859.6	7.58	8.81	31.68
rat195	BH + swap	3605.09	3767.97	55.19	62.2	56.5
(2323)	BH + insertion	3468.81	3749.51	49.32	61.41	42.68
	BH + reverse	3402.19	3591.48	46.46	54.61	55.54
	BH + s-reverse	3636.24	3833.81	56.53	65.04	61.11
	BH + NN + reverse	2582.33	2599.01	11.16	11.88	46.23

Table 1.
Experimental results of BH and its hybrids on the TSP datasets.

	Best deviation	Average deviation	CPU time
BH + swap	15.35	19.78	23.75
BH + insertion	12.42	18.5	20.70
BH + reverse	11.74	16.93	24.98
BH + swap-reverse	15.75	20.57	26.39
BH + NN + reverse	4.55	6.14	20.63

Table 2.
Mean of best, average deviations, and CPU time on the TSP datasets.

Figure 2 presents a set of obtained solutions by the hybrid black hole algorithm with a reverse operator on the benchmark test data sets.

Figure 3 shows the performance of algorithms on a sample of TSP datasets (berlin52-bier127). All algorithms, except BH + NN + reverse, are competitive results and get fairly higher performance when **Table 2** and **Figure 3** are compared. Although the BH + swap-reverse hybrid metaheuristic shows slightly worse performance, it is quite better at three datasets: eil76, pr76, and kroa100. On the other hand, the proposed approach gets slightly better performance in the mean of average deviations when **Table 2** and **Figure 3** are compared.

5. Conclusions and future work

In recent literature and engineering research of the field, solving discrete and engineering problems using hybrid heuristics and metaheuristics is an area of increasing interest. In this paper, the hybrid metaheuristics are experimented with the symmetric TSP instances. To evaluate the performance of hybrid metaheuristics, it has

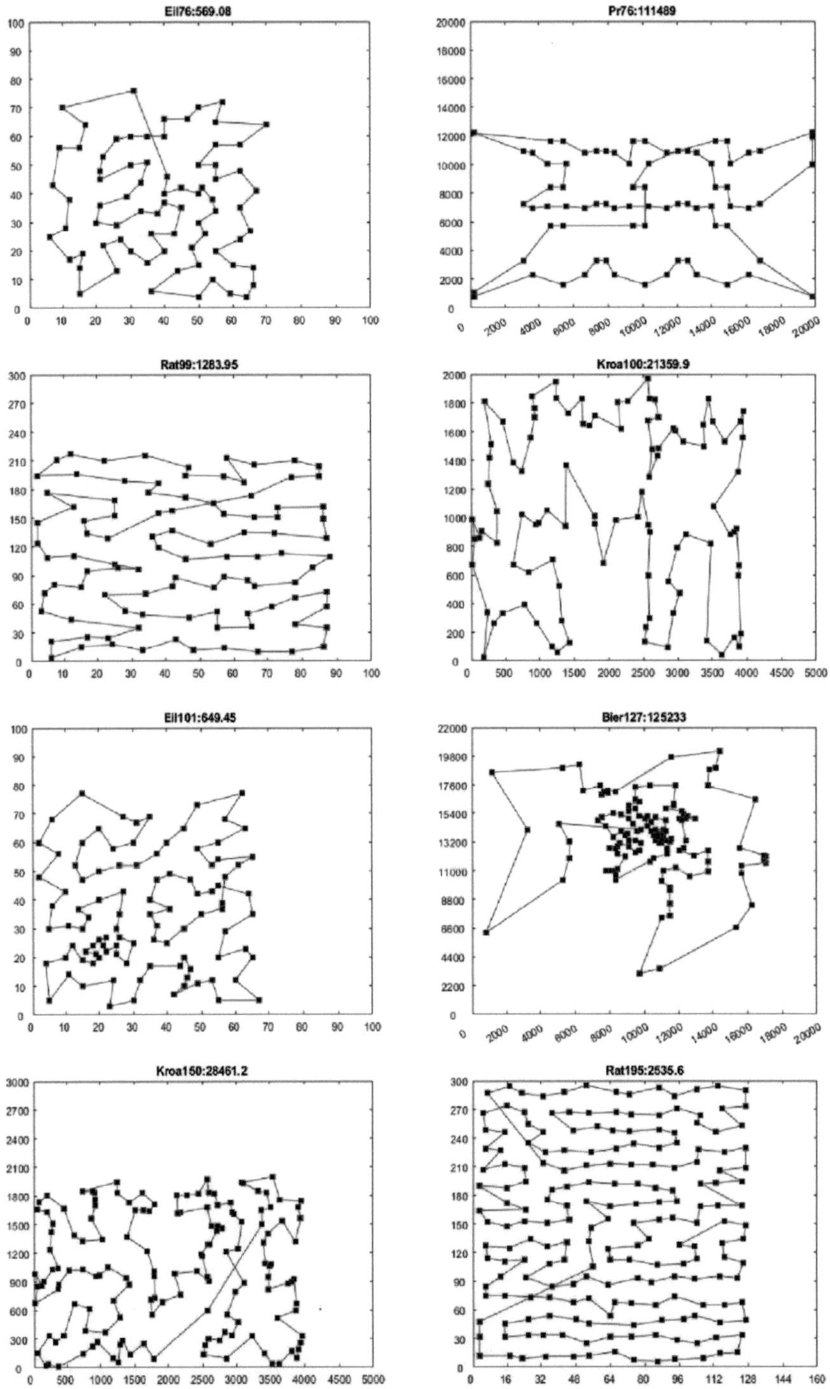

Figure 2.
A set of solutions found by the hybrid BH algorithm on the TSP datasets (black hole algorithm + nearest neighbor + reverse metaheuristic).

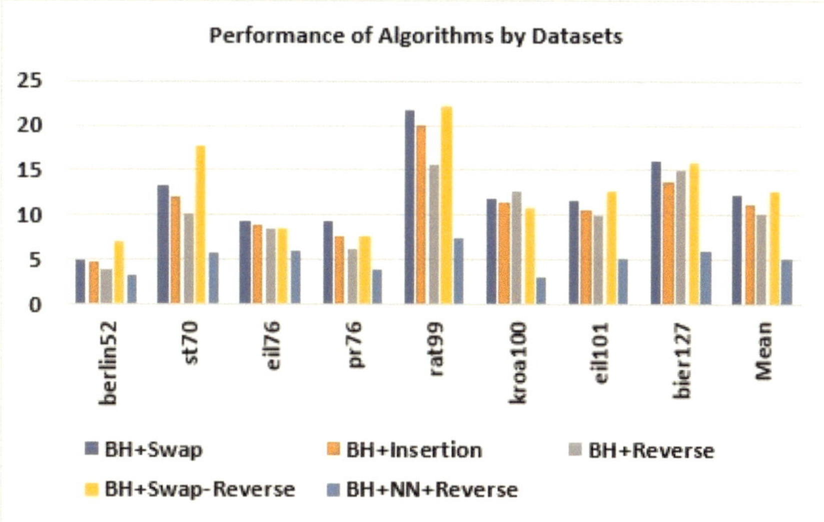

Figure 3.
Average deviations of hybrid algorithms on the TSP datasets.

been tested on ten medium-scale datasets. The experimental results show that the proposed hybrid algorithm BH + NN + reverse finds better solutions compared to the BH + swap, BH + insertion, BH + reverse, and BH + swap-reverse for all of the datasets and all of the optimal solutions. As CPU time is considered, the hybrid algorithm is considerably fast (20.63 secs.) to find near-optimal solutions. BH + reverse is the second algorithm when compared to the BH + swap, BH + insertion, and BH + swap-reverse. However, it needs a longer computational time (24.98) compared to BH + swap (23.75 sec.) and BH + insertion (20.70 secs.). It is a remarkable result that BH + insertion is the third algorithm to find near-optimal solutions. It is slightly slower than the proposed hybrid algorithm.

When detailed investigation is made on sample datasets (berlin52-bier127), the hybrid algorithms without the initial algorithm are quite challenging algorithms with each other. However, the proposed algorithm BH + NN + reverse is the best among the five alternatives. When the CPU time is calculated, the BH + insertion is the fastest algorithm (15.48 secs.), the proposed one is the second alternative (16.05 secs.), and BH + swap is the third alternative (16.58 secs). Nevertheless, the BH + reverse metaheuristic is the last alternative among hybrid algorithms (20.42 secs.).

In previous studies, the effect of heuristic algorithms on the performance of meta-heuristics has been strongly underlined. The exploration and exploitation phase of the algorithm directly depends on the type of heuristic being used. Those heuristics respond to different optimal solutions with different metaheuristics in longer iteration and computation time. In general, different heuristic algorithms contribute differently to the combinatorial and engineering problems. Conversely, the experimental results are valid for medium-sized TSP problems and a few number of iterations (1000–3000) in the application.

In future works, the hybrid metaheuristics can be further analyzed and combined with other heuristics and metaheuristics. Many applications could be done using those hybrid algorithms to evaluate performance analysis and solve engineering problems in applied mathematics, chemistry, biology, genetics, and nanotechnology. In another aspect, there could be application areas in industrial engineering, computer engineering, operations research, and combinatorial optimization, such as scheduling, assignment, timetabling, routing, location selection, inventory allocation, and logistics. To sum up, there are interdisciplinary working fields between science, technology, combinatorial optimization, and hybrid metaheuristic algorithms.

Acknowledgements

The author is responsible for all the manuscript, data availability, and coded optimization algorithms. The author declares that this work did not receive funding in the public, commercial, and non-profit sectors.

Conflict of interest

The author declares that there is no conflict of interest.

Thanks

The author thanks the editors and reviewers for their valuable comments and suggestions, which increased the clarity and scope of the article.

Author details

Mehmet Fatih Demiral
Department of Industrial Engineering, Burdur Mehmet Akif Ersoy University, Burdur, Turkey

*Address all correspondence to: mfdemiral@mehmetakif.edu.tr

References

[1] Dokeroglu T, Sevinc E, Kucukyilmaz T, Cosar A. A survey on new generation metaheuristic algorithms. Computers & Industrial Engineering. 2019;**137**:106040. DOI: 10.1016/j.cie.2019.106040

[2] Gohil NB, Dwivedi VV. A review on lion optimization: Nature inspired evolutionary algorithm. International Journal of Advanced in Management, Technology and Engineering Sciences. 2017;7:340-352. DOI: 16.10089. IJAMTES.2017.V7I12.15.2011

[3] Rajpurohit J, Sharma TK, Abraham A, Vaishali. Glossary of metaheuristic algorithms. International Journal of Computer Information Systems & Industrial Management Applications. 2017;**9**:181-205

[4] Khanra A, Maiti MK, Maiti M. Profit maximization of TSP through a hybrid algorithm. Computers & Industrial Engineering. 2015;**88**:229-236. DOI: 10.1016/j.cie.2015.06.018

[5] Tawhid MA, Savsani P. Discrete sine-cosine algorithm (dsca) with local search for solving traveling salesman problem. Arabian Journal for Science and Engineering. 2019;**44**:3669-3679. DOI: 10.1007/s13369-018-3617-0

[6] Sarhani M, Voß S, Jovanovic R. Initialization of metaheuristics: Comprehensive review, critical analysis, and research directions. International Transactions in Operational Research. 2023;**30**:3361-3397. DOI: 10.1111/itor.13237

[7] Fosin J, Carić T, Ivanjko E. Vehicle routing optimization using multiple local search improvements. Automatika. 2014;**55**:124-132. DOI: 10.7305/automatika.2014.01.580

[8] Contreras-Bolton C, Parada V. Automatic combination of operators in a genetic algorithm to solve the traveling salesman problem. PLoS One. 2015;**10**: e0137724. DOI: 10.1371/journal.pone.0137724

[9] Ter-Sarkisov A, Marsland S. K-bit-swap: A new operator for real-coded evolutionary algorithms. Soft Computing. 2017;**21**:6133-6142. DOI: 10.1007/s00500-016-2170-6

[10] Irnich S, Laganà D, Schlebusch C, Vocaturo F. Two-phase branch-and-cut for the mixed capacitated general routing problem. European Journal of Operational Research. 2015;**243**:17-29. DOI: 10.1016/j.ejor.2014.11.005

[11] Riera-Ledesma J, Salazar-González JJ. Solving school bus routing using the multiple vehicle traveling purchaser problem: A branch-and-cut approach. Computers & Operations Research. 2012;**39**:391-404. DOI: 10.1016/j.cor.2011.04015

[12] Riera-Ledesma J, Salazar-González JJ. A column generation approach for a school bus routing problem with resource constraints. Computers & Operations Research. 2013;**40**:566-583. DOI: 10.1016/j.cor.2012.08.011

[13] Yuan Y, Cattaruzza D, Ogier M, Semet F. A branch-and-cut algorithm for the generalized traveling salesman problem with time windows. European Journal of Operational Research. 2020; **286**:849-866. DOI: 10.1016/j.ejor.2020.04.024

[14] Aladag CH, Hocaoglu G, Basaran MA. The effect of neighborhood structures on tabu search algorithm in solving course timetabling problem.

Expert Systems with Applications. 2009; **36**:12349-12356. DOI: 10.1016/j.eswa.2009.04.051

[15] Kaspi M, Zofi M, Teller R. Maximizing the profit per unit time for the travelling salesman problem. Computers & Industrial Engineering. 2019;**135**:702-710. DOI: 10.1016/j.cie.2019.06.050

[16] Ali RS, Alnahwi FM, Abdullah AS. A modified camel travelling behavior algorithm for engineering applications. Australian Journal of Electrical and Electronics Engineering. 2019;**16**: 176-186. DOI: 10.1080/1448837X.2-019.1640010

[17] Bozorgi SM, Yazdani S. IWOA: An improved whale optimization algorithm for optimization problems. Journal of Computational Design and Engineering. 2019;**6**:243-259. DOI: 10.1016/j.jcde.2019.02.002

[18] Feng X, Liu Y, Yu H, Luo F. Physarum-energy optimization algorithm. Soft Computing. 2019;**23**:871-888. DOI: 10.1007/s00500-017-2796-z

[19] Yang XS. Nature-Inspired Metaheuristic Algorithms. 2nd ed. Frome: Luniver Press; 2010. p. 115

[20] Yildirim AE, Karci A. Applications of artificial atom algorithm to small-scale traveling salesman problems. Soft Computing. 2018;**22**:7619-7631. DOI: 10.1007/s00500-017-2735-z

[21] Yang Q, Chu SC, Pan JS, Chen CM. Sine cosine algorithm with multigroup and multistrategy for solving cvrp. Mathematical Problems in Engineering. 2020;**8184254**:10. DOI: 10.1155/2020/8184254

[22] Li Q, Ning H, Gong J, Li X, Dai B. A hybrid greedy sine cosine algorithm with differential evolution for global optimization and cylindricity error evaluation. Applied Artificial Intelligence. 2021;**35**:171-191. DOI: 10.1080/08839514.2020.1848276

[23] Haroun SA, Jamal B, Hicham EH. A performance comparison of GA and ACO applied to TSP. International Journal of Computer Applications. 2015; **117**:28-35. DOI: 10.5120/20674-3466

[24] TSPLIB. Discrete and Combinatorial Optimization [Internet]. 1995. Available from: http://comopt.ifi.uni-heidelberg.de/software/TSPLIB95/ [Accessed: January 15, 2023]

[25] Lin J, Zhou W, Wolfson O. Electric vehicle routing problem. Transportation Research Procedia. 2016;**12**:508-521. DOI: 10.1016/j.trpro.2016.02.007

[26] Szeto WY, Wu Y, Ho SC. An artificial bee colony algorithm for the capacitated vehicle routing problem. European Journal of Operational Research. 2011;**215**:126-135. DOI: 10.1016/j.ejor.2011.06.006

[27] Halim AH, Ismail I. Combinatorial optimization: Comparison of heuristic algorithms in travelling salesman problem. Archives of Computational Methods in Engineering. 2019;**26**: 367-380. DOI: 10.1007/s11831-017-9247-y

[28] Riera-Ledesma J, Salazar-González JJ. A heuristic approach for the travelling purchaser problem. European Journal of Operational Research. 2005; **162**:142-152

[29] Kucukoglu I. The traveling purchaser problem with fast service option. Computers & Operations Research. 2022;**141**:105700. DOI: 10.1016/j.cor.2022.105700

[30] Bektas T. The multiple traveling salesman problem: An overview of formulations and solution procedures. Omega. 2006;**34**:209-219

[31] Hatamlou A. Black hole: A new heuristic optimization approach for data clustering. Information Sciences. 2013; **222**:175-184. DOI: 10.1016/j.ins.2012.08.02 3

[32] Kumar S, Datta D, Singh SK. Black hole algorithm and its applications. In: Azar A, Vaidyanathan S, editors. Computational Intelligence Applications in Modeling and Control. Studies in Computational Intelligence. Cham: Springer; 2015. pp. 147-170. DOI: 10.1007/978-3-319-11017-2_7

[33] Ágota B, Tamás B, Béla I. Optimization of consignment-store-based supply chain with black hole algorithm. Complexity. 2017;**6038973**: 12. DOI: 10.1155/2017/6038973

[34] Hatamlou A. Solving travelling salesman problem using black hole algorithm. Soft Computing. 2018;**22**: 8167-8175. DOI: 10.1007/s00500-017-2760-y

[35] Kanagasabai L. Amplified black hole algorithm for real power loss reduction. International Journal of Research in Industrial Engineering. 2020;**9**:130-142. DOI: 10.22105/riej.2020.214468.1114

[36] Demiral MF. A performance comparison of hybrid black hole algorithm with popular meta-heuristics in traveling salesman problem. In: Ceylan AC, Kaya CM, editors. Research & Reviews in Social, Human and Administrative Sciences. 1st ed. Ankara: Gece Publishing; 2022. pp. 217-232